ENCYCLOPAEDIA OF ENVIRONMENTAL BIOTECHNOLOGY

ENCYCLOPAEDIA OF ENVIRONMENTAL BIOTECHNOLOGY

Vol. 3

Shahzad Ahmad

ANMOL PUBLICATIONS PVT. LTD.
NEW DELHI - 110 002 (INDIA)

ANMOL PUBLICATIONS PVT. LTD.
H.O.: 4374/4B, Ansari Road, Darya Ganj,
New Delhi-110 002 (India)
Ph.: 23278000, 23261597
B.O.: No. 1015, Ist Main Road, BSK IIIrd Stage
IIIrd Phase, IIIrd Block,
Bangalore - 560 085 (India)
Visit us at: www.anmolpublications.com

Encyclopaedia of Environmental Biotechnology

First Published, 2008

ISBN 978-81-261-3539-4 (Set)

PRINTED IN INDIA

Printed at Mehra Offset Press, Delhi.

Contents

Preface

Environmental biotechnology is the applied technology for the study of natural environment. Environmental biotechnology could also imply that one tries to harness biological process for commercial use and exploitation. Environmental biotechnology, by contrast, deals readily and apparently with dramatic topics, and though their importance, albeit different, may be every bit as great as their direct relevance is quite readily appreciated by the bulk of population. Cleaning up contamination and dealing rationally with waste is of course, in everybody's best interests, but for most people, this is like simply addressing a problem. Some of the potentially beneficial uses of biological engineering, which touch the lives of the majority of people, however directly involve much simpler approaches. In many respects, there is a logical link between the aims of environmental biotechnology.

Environmental biotechnology is fundamentally rooted in wastes, in its various guises, typically being concerned with the remediation of the contamination caused by previous use and the impact reduction of current activity or the control of pollution. Thus, the principal aim of this stream are the manufacture of products in environmentally harmonious manner, which allow for the minimisation of harmful solids, liquids or gaseous outputs or clean up the residual effects of earlier human occupations. Environmental biotechnologists may enhance or optimise conditions for existing biological systems to make their activities happen faster or more efficiently or they resort to some form of alteration to bring about the desired results. The variety of organism, which may play a part in environmental applications of biotechnology is wide and ranging from microbes through trees and all are utilised on one of the same three fundamental bases.

Accordingly, the range of businesses to which environmental biotechnology has potential relevance is almost limitless. All commercial operations generate wastes in one form or the other. Additionally, biotechnology can be of further commercial significance by converting low-cost organic feed stocks into high-value products.

Biotechnology based methods are seen as essential to help meet European Union (EU) targets for biowaste diversion and reductions in pollutants. Across the world, the existing regulations on environmental pollution are required to be rigorously enforced, with more stringent compliance standards implemented. All of this is expected to stimulate the sale of biotechnology based environmental processing methods significantly and in particular, the global market share is projected to grow faster than the general biotic sector trends.

This exhaustive work has a perfect treatment with environmental biotechnology in an objective manner. Expectedly, this work would be capable to create a sort of awareness among the people and make the study of environmental biotechnology easier and more meaningful.

Editor

1

Soil Remediation

The most common soil contaminants are petroleum-based. Hydrocarbons from diesel fuel and gasoline are widespread problems, as are polyaromatic hydrocarbons (PAH). Many PAHs are known carcinogens, and others are suspected problem chemicals that need to be kept from contaminating drinking water. Chemicals tend to spread through soil by diffusion and convection. Diffusion is a molecular transport that is motivated by differences in concentration. Chemicals are a lot like people, in that if you put a bunch of them in one area but there aren't many in the surrounding region they'll slowly sprawl out into the unpopulated region. Convection is a means of molecular motion where the driving force is provided by a fluid such as rain or wind. Water will pick up pollutant particles as it soaks into the earth and carry them further from the initial spill. Fortunately, the mechanisms that spread pollutants can also be used to remove them.

There are Two Distinct Classes of Soil Remediation: *In situ*, or on-site, and *ex situ*, or off- site. On-site cleanups are often preferred because they are cheaper. On the other hand, excavating a contaminated area and transporting it to a remote site before cleaning it can often be more complete. *Ex situ* remediation also has the added bonus of taking the bulk of contaminants off-site before they can spread further. In addition, *in situ* situations are limited because only the topside of the soil is accessible. Some drilling can be done, but ultimately the contaminants must fight gravity and surface. These environmental limitations force *in situ* remediation to fall into three categories: washing, venting, and bioremediation. Off-site facilities have the luxury of more complete control over the cleaning chemical processes. Soil can be isolated in a controlled environment, without weather problems. More caustic chemicals can be used and acid leaching can be conducted *ex situ* without harming the environment. Soil is often put into a sealed dome, and cleansing fluids are pumped in while wastes are pumped out and treated.

On-site Soil Cleaning Typically Employs Two Methods: The transport of the pollutants out of the ground in a safe manner or the chemical degradation of pollutants. In "washing," fluids like water or other chemicals are pumped in to force the pollutants out of the ground. Liquids that dissolve the pollutant are chosen. Many pollutants, however, aren't easily solvated, or are only soluble in caustic chemicals, limiting the applicability of *in situ* washing. Sometimes a trick from enhanced oil recovery is used to flush out less soluble compounds. Before an oil well is completely dry, the oil needs to be forced out by water and then a viscous polymer solution. Similarly, viscous fluids or foams can be pumped through the soil to collect the chemicals. Typically foams are pumped into the ground through one hole and the contaminated residue is pumped out from another hole on the other side of the contaminated region. Since soil contamination is usually a top-soil problem, and oil recovery is performed in deeper, denser rocks, the fluid pressure used to remove pollutants is much less than that used in enhanced oil recovery.

The mechanism for soil venting is similar to that of

washing, except it employs gases instead of liquid to transport the pollutant into the collector. Some chemicals are fairly volatile and can be collected by pumping air into pilot holes while sucking it out of other holes. Soil density has an obvious influence on the distance between the holes that are effective and economical. Warm air and heating coils can be used to increase the vapour pressure of a compound. This increases the amount that can be transported within a given time. Chemical transport can also be motivated by electrical fields. An anode and cathode can be inserted into contaminated ground and used to create an electrical field. Since contaminants may be polar and have a slight charge, they are be pulled by electro-magnetic forces towards the terminal with an opposite charge. After sufficient time has passed the terminal and its immediately surrounding soil are extracted and treated off-site.

Other chemicals can be decomposed into harmless gases. This could happen naturally as bacteria break complex chemicals into CO_2 and smaller parts. Unfortunately this is a fairly slow process, but it can be speed up by adding oxygen or nutrients into the environment for the bacteria. It still takes usually one to two years. Another mode of degradation is the addition of chemicals. Ozone has been show to break down polyaromatic hydrocarbons. It can be pumped into the ground and will degrade pollutants as it passes through the soil. There are many methods to remove pollutants from the soil, but they don't all work for all chemicals. This makes removal a delicate process. The method must be tailored to the individual pollutant, the soil type, and the surrounding geology. Cost is also a factor in selecting a method of remediation. It may be that increasing preventative measures is the best answer to soil remediation problems.

At the dawn of the 21st century, biotechnology is emerging as a key enabling technology for sustainable environmental protection and stewardship. Biotechnology offers a state-of-the-art account of environmental biotechnology both in emerging and in more mature technological applications of soil remediation and cleanup of contaminated sites. Harnessing the potential of microorganisms and plants as eco-efficient and robust cleanup agents in a variety of practical situations is not only possible but is becoming widespread practice.

Thermal Depolymerisation

Thermal depolymerisation (TDP) is a process for the reduction of complex organic materials (usually waste products of various sorts, often known as biomass and plastic) into light crude oil. It mimics the natural geological processes thought to be involved in the production of fossil fuels. Under pressure and heat, long chain polymers of hydrogen, oxygen, and carbon decompose into short-chain petroleum hydrocarbons with a maximum length of around 18 carbons.

Similar Processes

Thermal depolymerisation is sometimes mistaken for similar processes:

- *Thermochemical conversion (TCC)* is limited to the changing of manure to crude oil.
- *Thermal conversion process (TCP)* is limited to the changing of manure and vegetable waste to crude oil.

Thermal depolymerisation can change many carbon-based materials into crude oil and methane, and is not limited to manure or vegetable waste.

History

Thermal depolymerisation is similar to the geological processes that produced the fossil fuels used today, except that the technological process occurs in a timeframe measured in hours. Until recently, the human-designed processes were not efficient enough to serve as a practical source of fuel—more energy was required than was produced.

Many previous methods which create hydrocarbons

through depolymerisation used dry materials (or anhydrous pyrolysis), which requires expending a lot of energy to remove water. However, there has been work done on hydrous pyrolysis methods, in which the depolymerisation takes place with the materials in water. In US patent 2,177,557, issued in 1939, Bergstrom and Cederquist discuss a method for obtaining oil from wood in which the wood is heated under pressure in water with a significant amount of calcium hydroxide added to the mixture. In the early 1970's Herbert R. Appell and coworkers worked with hydrous pyrolysis methods, as exemplified by US patent 3,733,255 (issued in 1973), which discusses the production of oil from sewer sludge and municipal refuse by heating the material in water, under pressure, and in the presence of carbon monoxide.

An approach that exceeded break-even was developed by Illinois microbiologist Paul Baskis in the 1980s and refined over the next 15 years. The technology was finally developed for commercial use in 1996 by Changing World Technologies (CWT). Brian S. Appel (CEO of CWT) took the technology in 2001 and expanded and changed it into what is now referred to as TCP and has applied for several patents. A Thermal Depolymerisation demonstration plant was completed in 1999 in Philadelphia by Thermal Depolymerisation, LLC, and the first full-scale commercial plant was constructed in Carthage, Missouri, about 100 yards (100 m) from ConAgra Foods' massive Butterball turkey plant, where it is expected to process about 200 tons of turkey waste into 500 barrels (21,000 US gallons or 80 m^3) of oil per day.

Theory and Process

In the method used by CWT, the water improves the heating process and contributes hydrogen to the reactions.

The feedstock material is first ground into small chunks, and mixed with water if it is especially dry. It is then fed into a reaction chamber where it is heated to around 250°C and subjected to 600 psi (4 MPa) for approximately 15 minutes, after which the pressure is rapidly released to boil off most of the water. The result is a mix of crude hydrocarbons and solid minerals, which are separated out. The hydrocarbons are sent to a second-stage reactor where they are heated to 500°C, further breaking down the longer chains, and the resulting mix of hydrocarbons is then distilled in a manner similar to conventional oil refining.

Working with turkey offal as the feedstock, the process proved to have yield efficiencies of approximately 85 per cent; in other words, the energy contained in the end products of the process is 85 per cent of the energy contained in the inputs to the process (most notably the energy content of the feedstock, but also including electricity for pumps and natural gas for heating). Alternatively, if one considers the energy content of the feedstock to be free (i.e., waste material from some other process), one could consider the energy efficiency of the process to be 560 per cent (85 units of energy made available for 15 units of energy consumed).

The company claims that 15 to 20 per cent of feedstock energy is used to provide energy for the plant. The remaining energy is available in the converted product. Higher efficiencies may be possible with drier and more carbon-rich feedstocks, such as waste plastic.

By comparison, the current processes used to produce ethanol and biodiesel from agricultural sources have energy efficiencies in the 320 per cent range when the energy used to produce the feedstocks is considered (in this case, usually sugar cane, corn, soybeans and the like). As these energy efficiencies include the energy cost to produce the feedstock and the above TDP energy efficiency does not, these values are not directly comparable.

The process breaks down almost all materials that are fed into it. TDP even efficiently breaks down many types of hazardous materials, such as poisons and difficult-to-destroy biological agents such as prions.

Feedstocks and Outputs with Thermal Depolymerisation

Feedstock	*Output*		*Feedstock*	*Output*		*Feedstock*	*Output*	
Plastic bottle	Oil	70	Turkey offal	Oil	39%	Paper (cellulose)	Oil	8%
	Gas	16%		Gas	6%		Gas	48%
	Carbon solids	6%		Carbon solids	5%		Carbon solids	24%
Sewage sludge	Gas	9%	Medical waste	Gas	10%	Tires	Gas	10%
	Carbon solids	8%		Carbon solids	5%		Carbon and metal solids	42%
	Water	57%		Water	20%		Water	4%

Carthage Plant Products

The yield from one ton (907kg) of turkey waste is 600 pounds (272 kg) petroleum, 100 pounds (45 kg) butane/ methane, and 60 pounds (27kg) minerals. In addition, the water is recycled back into the system for reuse.

The Carthage, MO plant produces API 40+, a high value crude oil comparable to diesel fuel. It contains light and heavy naphthas, a kerosene, and a gas oil fraction, with essentially no heavy fuel oils, tars, asphaltenes, or waxes present.

Classification of TDP-40 Oil by PONA

PONA	*wt%, D-5443 method*
Paraffins	22 %
Olefins	14 %
Naphthenes	3 %
Aromatics	6 %
C14/C14+	55 %
TOTAL	100 %

The fixed carbon solids produced by the TDP process have multiple uses as a filter, a fuel source and a fertilizer. It can be used as activated carbon in wastewater treatment, as a fertilizer, or as a fuel similar to coal.

Advantages

The process can break down organic poisons, due to

breaking chemical bonds and destroying the molecular shape needed for the poison's activity. It is highly effective at killing pathogens, including prions. It can also safely remove heavy metals from the samples by converting them from their ionised or organometallic forms to their stable oxides which can be safely separated from the other products.

Potential Sources of Waste Inputs

The Environmental Protection Agency estimates that in 2001, there were 229 million tons of municipal solid waste, or 4.4 pounds generated per day per person in the USA. Industrial facilities in the USA create 7.6 billion tons of industrial wastes each year, however 97 per cent of that waste is water, which means only 228 million tons of potential feedstock remains. In failing to mention the latter, Changing World Technologies tend to overestimate the potential benefit that the country may reap from wide-scale implementation of the process.

Limitations

The process only breaks long molecular chains into shorter ones, so small molecules such as carbon dioxide or methane cannot be converted to oil through this process. Neither can thermal depolymerisation be used to remove radioactivity from radioactive waste.

Many agricultural and animal wastes could be processed, but many of these are already used as fertilizer, animal feed, and, in some cases, as feedstocks for paper mills or as boiler fuel.

Price and Design Changes

Reports from 2005 summarised some economic setbacks which the Carthage plant encountered since its planning stages. It was thought that concern over mad cow disease would prevent the use of turkey waste and other animal products as cattle feed, and thus this waste would be free. As it turns out, turkey waste may still be used as feed in the United States, so that the facility must purchase that feed stock at a cost of $30 to $40 per ton, adding $15 to $20 per barrel to the cost of the oil. Final cost, as of January 2005, was $80/barrel ($1.90/gal).

The above cost of production also excludes the operating cost of the thermal oxidiser and scrubber added in May 2005 in response to odour complaints.

A biofuel tax credit of roughly $1 per US gallon on production costs was not available because the oil produced did not meet the definition of "biodiesel" according to the relevant American tax legislation. The Energy Policy Act of 2005 specifically added thermal depolymerisation to a $1 renewable diesel credit, which became effective at the end of 2005.

Company Expansion

The company has explored expansion in California, Pennsylvania, and Virginia, and is presently examining projects in Europe, where animal products cannot be used as cattle feed. TDP is also being considered as an alternative means for sewage treatment in the United States.

The pilot plant in Carthage was temporarily shut down due to smell complaints. It was soon restarted when it was discovered that few of the odours were generated by the plant. Furthermore, the plant agreed to install an enhanced thermal oxidiser and to upgrade its air scrubber system under a court order. Since the plant is located only four blocks from the tourist-attracting town centre, this has strained relations with the mayor and citizens of Carthage.

According to a company spokeswoman, the plant has received complaints even on days when it is not operating. She also contended that the odours may not have been produced by their facility, which is located near several other agricultural processing plants.

In December 29, 2005, the plant was ordered by the state governor to shut down once again over allegations of foul odours as reported by MSNBC.

As of March 7, 2006, the plant has begun limited test runs to validate it has resolved the odour issue.

As of August 24, 2006, the last lawsuit connected with the odour issue has been dismissed and the problem is acknowledged as fixed. In late November, however, another complaint was filed over bad smells. This complaint was closed on January 11th of 2007 with no fines assessed.

Thermal Treatment: Ex Situ

Ex situ thermal treatment generally involves the destruction or removal of contaminants through exposure to high temperature in treatment cells, combustion chambers, or other means used to contain the contaminated media during the remediation process. The main advantage of *ex situ* treatments is that they generally require shorter time periods, and there is more certainty about the uniformity of treatment because of the ability to screen, homogenise, and continuously mix the contaminated media; however, *ex situ* processes require excavation of soils, which increases costs and engineering for equipment, permitting, and materials handling worker safety issues.

Thermal processes use heat to separate, destroy, or immobilise contaminants. Thermal desorption and hot gas decontamination are separation technologies. Pyrolysis and conventional incineration destroy the contaminants. Vitrification destroys or separates organics and immobilises some inorganics.

Incineration is a heat-based technology that has been used for many years to burn and destroy contaminated materials.

Ex Situ Thermal Desorption involves the application of heat to excavated wastes to volatilise organic contaminants and water. Typically, a carrier gas or vacuum system transports the volatilised water and organics to a treatment system, such as a thermal oxidation or recovery unit. Based on the operating temperature of the desorber, thermal desorption processes can be categorised as either high-temperature thermal desorption

(320 to 560°C or 600 to 1,000°F) or low-temperature thermal desorption (90 to 320°C or 200 to 600°F).

Hot Gas Decontamination involves raising the temperature of contaminated solid material or equipment to 260°C (500°F) for a specified period of time. The gas effluent from the material is treated in an afterburner system to destroy all volatilised contaminants. This method will permit reuse or disposal of scrap as non-hazardous material.

Plasma High-temperature Recovery uses a thermal treatment process applied to solids and soils that purges contaminants as metal fumes and organic vapours. The vapours can be burned as fuel, and the metals can be recovered and recycled.

Pyrolysis is defined as chemical decomposition induced in organic materials by heat in the absence of oxygen. Pyrolysis typically occurs under pressure and at operating temperatures above 430°C (800°F). The pyrolysis gases require further treatment. The target contaminant groups for pyrolysis are SVOCs and pesticides. The process is applicable for the separation of organics from refinery wastes, coal tar wastes, wood-treating wastes, creosote-contaminated soils, hydrocarbon-contaminated soils, mixed (radioactive and hazardous) wastes, synthetic rubber processing wastes, and paint waste.

Thermal Offgas Treatment is one of several approaches that can be used to cleanse the offgases generated from primary treatment technologies, such as air stripping and soil vapour extraction. In addition to the established thermal treatments, organic contaminants in gaseous form can be destroyed using innovative or emerging technologies, such as alkali bed reactors.

Thermal Desorption

Thermal desorption is an environmental remediation technology that utilises heat to increase the volatility of contaminants such that they can be removed (separated) from the solid matrix (typically soil, sludge or filter cake).

Thermal desorption is not incineration. The volatilised contaminants are then either collected or thermally destroyed. A thermal desorption system therefore has two major components; the desorber itself and the offgas treatment system.

Thermal desorption first appeared as an environmental treatment technology in 1985 when it was specified in the Record of Decision (ROD) for the McKin site in Maine. The latest status report from the United States Environmental Protection Agency (USEPA) shows that thermal desorption has been used at 69 Superfund sites through FY2000. In addition, hundreds of remediation project have been completed using thermal desorption at non-Superfund sites.

For *ex situ* on-site treatment options, only incineration and stabilisation has been used at more Superfund sites. Incineration suffers from poor public acceptance. Stabilisation does not provide a permanent remedy, since the contaminants are still on-site. Thermal desorption is a widely accepted technology that provides a permanent solution at an economically competitive cost.

Desorbers

Numerous desorber types are available today. Some of the more common types are listed below.

Indirect Fired Rotary, Direct Fired Rotary, Heated Screw (Hot Oil, Molten Salt, Electric) Infrared Microwave.

Most indirect fired rotary systems use an inclined rotating metallic cylinder to heat the feed material. The heat transfer mechanism is usually conduction through the cylinder wall. In this type of system neither the flame nor the products of combustion can contact the feed solids or the offgas. Think of it as a rotating pipe inside a furnace with both ends sticking outside of the furnace. The cylinder for full scale transportable systems is typically five to eight feet in diameter with heated lengths ranging from twenty to fifty feet. With a carbon steel shell, the maximum solids temperature is around 1,000°F, while temperatures of 1,800°F with special alloy cylinders are attainable. Total residence time in this type of desorber normally ranges from 30 to 120 minutes. Treatment capacities can range from 2 to 30 tons per hour for transportable units.

Direct fired rotary desorbers have been used extensively over the years for petroleum contaminated soils and soils contaminated with RCRA hazardous wastes as defined by the USEPA. A 1992 paper on treating petroleum contaminated soils estimated that between 20 and 30 contractors have 40 to 60 rotary dryer systems available. Today, it is probably closer to 6 to 10 contractors with 15 to 20 portable systems commercially available. The majority of these systems utilise a secondary combustion chamber (afterburner) or catalytic oxidiser to thermally destroy the volatilised organics. A few of these systems also have a quench and scrubber after the oxidiser which allows them to treat soils containing chlorinated organics such as solvents and pesticides. The desorbing cylinder for full scale transportable systems is typically four to ten feet in diameter with heated lengths ranging from twenty to fifty feet. The maximum practical solids temperature for these systems is around 750 to 900°F depending on the material of construction of the cylinder. Total residence time in this type of desorber normally ranges from 3 to 15 minutes. Treatment capacities can range from 6 to over 100 tons per hour for transportable units.

Heated screw systems are also an indirect heated system. Typically they use a jacketed trough with a double auger that intermeshes. The augers themselves frequently contain passages for the heating medium to increase the heat transfer surface area. Some systems use electric resistance heaters instead of a heat transfer media and may employ a single auger in each housing. The augers can range from 12 to 36 inches in diameter for full scale systems, with lengths up to 20 feet. The auger/trough assemblies can be connected in parallel and/or series to increase throughput. Full scale capabilities up to 4 tons per hour have been demonstrated. This type of system has been most successful treating refinery wastes.

In the early days, there was a continuous infrared system that is no longer in common use. In theory, microwaves would be an excellent technical choice since uniform and accurately controlled heating can be achieved with no heat transfer surface fouling problems. One can only guess that capital and/or energy costs have prevented the development of a microwave thermal desorber at the commercial scale.

Offgas Treatment

There are only three basic options for offgas treatment available. The volatilised contaminants in the offgas can either be discharge to atmosphere, collected or destroyed. In some cases, both a collection and destruction system are employed. In addition to managing the volatilised components, the particulate solids (dust) that exit the desorber must also be removed from the offgas.

When a collection system is used, the offgas must be cooled to condense the bulk of the volatilised components into a liquid. The offgas will exit most desorbers in the 350-900°F range. The offgas is then typically cooled to somewhere between 120 and 40°F to condense the bulk of the volatilised water and organic contaminants. Even at 40°F, there may be measurable amounts of non-condensed organics. For this reason, after the condensation step, further treatment of the offgas is usually required. The cooled offgas may be treated by carbon adsorption, or thermal oxidation. Thermal oxidation can be accomplished using a catalytic oxidiser, an afterburner or by routing the offgas to the combustion heat source for the desorber. The volume of gas requiring treatment for indirect fired desorbers is a fraction of that required for a direct fired desorber. This requires smaller air pollution control trains for the gaseous process vent emissions. Some thermal desorption systems recycle the carrier gas, thereby further reducing the volume of gaseous emissions.

The condensed liquid from cooling the offgas is separated into organic and aqueous fractions. The water is either disposed of or used to cool the treated solids and prevent dusting. The condensed liquid organic is removed from the site. Depending on its composition, the liquid is either recycled as a supplemental fuel or destroyed in a fixed base incinerator. A thermal desorber removing 500 mg/kg of organic contaminants from 20,000 tons of soil will produce less than 3,000 gallons of liquid organic. In essence 20,000 tons of contaminated soil could be reduced to less than one tank truck of extracted liquid residue for off-site disposal.

Desorbers using offgas destruction systems use combustion to thermally destroy the volatilised organics components forming CO, CO_2, NOx, SOx and HCl. The destruction unit may be called an afterburner, secondary combustion chamber, or thermal oxidiser. Catalytic oxidisers may also be used if the organic halide content of the contaminated media is low enough. Regardless of the name, the destruction unit is used to thermally destroy the hazardous organic constituents that were removed (volatilised) from the soil or waste.

What Does it Do?

Thermal desorption systems are designed to remove organic compounds from solid matrices such as soils, sludges and filter cakes without thermally destroying them. It is a separation technology, not a destruction technology. Since it is a thermal process, there is a common belief that temperature is the only significant parameter to monitor. While it is true that better removal efficiencies are usually achieved at higher temperatures, other factors must be considered. Since the process is governed by mass transfer, heating time and the amount of mixing are also key parameters in optimising removal efficiency.

Thermal desorption had successfully been used for just about every organic contaminant found to date. It has also been used to remove mercury. This does not mean that thermal desorption is the best choice for every contaminated soil project. With the exception of mercury, thermal desorption can not be used to remove heavy metals. However; stabilising agents can be added to the feed soil or the treated soil during the rewetting step to decrease the leachability of most metals.

Most thermal desorbers can mechanically treat most any moisture level, but at higher moisture levels the cost can become prohibitive. Excessive moisture in the feed soil must be avoided. As an example, a small direct fired system that could process 15 tons per hour with 10 per cent moisture in the feed would drop to 10.5 tons per hour if the moisture increased to 18 per cent. This 30 per cent decrease in process rate would result in a 40 per cent increase in the treatment cost per ton. For most projects, if the feed material contains more than 25 per cent moisture dewatering or drying options should be investigated.

Just about any media can be treated with thermal desorption to remove organic contaminants and mercury. Moisture and project size are the two factors that most dramatically affect treatment cost. Thermal desorption will work, but is it the correct choice for your site.

How clean will the soil be? As a rule of thumb, most contaminants can be reduced to less than 1.0 mg/kg (ppm) if the starting level is less than 500 mg/kg. Another way to make preliminary estimate is to assume a removal efficiency (not to be confused with DRE) of 98 to 99.99 per cent. Most volatile organic compounds and solvents can easily be reduced to less than 0.1 mg/kg and frequently to below 0.01 mg/kg. Most semi-volatile compounds can be reduced to less than 1.0 mg/kg and frequently to less than 0.1 mg/kg. The various PCB Aroclors can be reduced to less than 2.0 mg/kg and frequently to less than 0.5 mg/kg. Most pesticides will have a 99.5 to 99.99 per cent removal.

The treated soil is usually backfilled in to the excavation. The material compacts well and most compaction specifications can be met. Since the treated soil is essentially sterile, re-growth can be accomplished by covering with top soil or adding nutrients to the treated soil such as fertilizer or biosolids.

What Does it Cost?

Where suitable, direct fired thermal desorbers with afterburners will cost less than an indirect desorber with offgas collection. The indirect desorber will usually be less costly than on-site incineration. In many cases, thermal desorption is less costly than hog-and-haul options. Projects vary widely in size, type of thermal system to use, moisture level and treatment standards that price ranges do not make sense, since they cover almost one order of magnitude.

Ex Situ Physical/Chemical Treatment for Soil, Sediment, Bedrock and Sludge

The main advantage of *ex situ* treatment is that it generally requires shorter time periods than *in situ* treatment, and there is more certainty about the uniformity of treatment because of the ability to homogenise, screen, and continuously mix the soil. *Ex situ* treatment, however, requires excavation of soils, leading to increased costs and engineering for equipment, possible permitting, and material handling/worker exposure conditions.

Physical/chemical treatment uses the physical properties of the contaminants or the contaminated medium to destroy (i.e., chemically convert), separate, or immobilise the contamination. Chemical reduction/oxidation and dehalogenation (APEG, BCD or glycolate) are destruction technologies. Soil washing, SVE, and solvent extraction are separation techniques, and Solidification/Stabilisation (S/S) is an immobilisation technique.

Physical/chemical treatment is typically cost-effective and can be completed in short time periods (in comparison with biological treatment). Equipment is readily available and is not engineering or energy-intensive. Treatment residuals from separation techniques will require treatment or disposal, which will add to the total project costs and may require permits.

Available *ex situ* physical/chemical treatment technologies include chemical extraction, chemical reduction/oxidation, dehalogenation (APEG, BCD or glycolate), separation, soil washing, and solidification/stabilisation.

Bioventing is a promising new technology that stimulates the natural *in situ* biodegradation of any aerobically degradable compounds in soil by providing oxygen to existing soil microorganisms. In contrast to soil vapour vacuum extraction, bioventing uses low air flow rates to provide only enough oxygen to sustain microbial activity. Oxygen is most commonly supplied through direct air injection into residual contamination in soil. In addition to degradation of adsorbed fuel residuals, volatile compounds are biodegraded as vapours move slowly through biologically active soil.

Enhanced bioremediation is a process in which indigenous or inoculated microorganisms (e.g., fungi, bacteria, and other microbes) degrade (metabolise) organic contaminants found in soil and/or ground water, converting them to innocuous end products. Nutrients, oxygen, or other amendments may be used to enhance bioremediation and contaminant desorption from subsurface materials.

Aerobic

In the presence of sufficient oxygen (aerobic conditions), and other nutrient elements, microorganisms will ultimately convert many organic contaminants to carbon dioxide, water, and microbial cell mass.

Enhanced bioremediation of soil typically involves the percolation or injection of ground water or uncontaminated water mixed with nutrients and saturated with dissolved oxygen. Sometimes acclimated microorganisms (bioaugmentation) and/or another oxygen source such as hydrogen peroxide are also added. An infiltration gallery or spray irrigation is typically used for shallow contaminated soils, and injection wells are used for deeper contaminated soils.

Although successful *in situ* bioremediation has been

demonstrated in cold weather climate, low temperature slows the remediation process. For contaminated sites with low soil temperature, heat blankets may be used to cover the soil surface to increase the soil temperature and the degradation rate.

Enhanced bioremediation may be classified as a long-term technology which may take several years for cleanup of a plume.

Anaerobic

In the absence of oxygen (anaerobic conditions), the organic contaminants will be ultimately metabolised to methane, limited amounts of carbon dioxide, and trace amounts of hydrogen gas. Under sulphate-reduction conditions, sulphate is converted to sulphide or elemental sulphur, and under nitrate-reduction conditions, dinitrogen gas is ultimately produced.

Sometimes contaminants may be degraded to intermediate or final products that may be less, equally, or more hazardous than the original contaminant. For example, TCE anaerobically biodegrades to the persistent and more toxic vinyl chloride. To avoid such problems, most bioremediation projects are conducted *in situ*. Vinyl chloride can easily be broken down further if aerobic conditions are created.

White Rot Fungus

White rot fungus has been reported to degrade a wide variety of organopollutants because of its lignin-degrading or wood-rotting enzymes. Two different treatment configurations have been tested for white rot fungus, *in situ* and bioreactor. An aerobic system using moisturised air on wood chips is used in a reactor for biodegradation. A reactor was used in the bench-scale trial of the process. In the pilot-scale project, an adjustable shredder was used for making chips for the open system. The open system is similar to composting, with wood chips on a liner or hard contained surface that is covered. Temperature is not controlled in this type of system. The optimum temperature for biodegradation with lignin-degrading fungus ranges from 30 to 38°C (86 to 100°F). The heat of the biodegradation reaction will help to maintain the temperature of the process near the optimum.

Although white rot fungus degradation of TNT has been reported in laboratory-scale settings using pure cultures, several factors increase the difficulty of using this technology for full-scale remediation, and it has not yet been proven successful at this level. These factors include competition from native bacterial populations, toxicity inhibition, chemical sorption, and the inability to meet risk-based cleanup levels. White rot works best in nitrogen-limited environments.

Phytoremediation is a process that uses plants to remove, transfer, stabilise, and destroy contaminants in soil and sediment. The mechanisms of phytoremediation include enhanced rhizosphere biodegradation, phytoextraction (also called phytoaccumulation), phytodegradation, and phytostabilisation.

Enhanced Rhizosphere Biodegradation: Enhanced rhizosphere biodegradation takes place in the soil immediately surrounding plant roots. Natural substances released by plant roots supply nutrients to microorganisms, which enhances their biological activities. Plant roots also loosen the soil and then die, leaving paths for transport of water and aeration. This process tends to pull water to the surface zone and dry the lower saturated zones.

The most commonly used flora in phytoremediation projects are poplar trees, primarily because the trees are fastgrowing and can survive in a broad range of climates. In addition, poplar trees can draw large amounts of water (relative to other plant species) as it passes through soil or directly from an aquifer. This may draw greater amounts of dissolved pollutants from contaminated media and reduce the amount of water that may pass through soil or an aquifer, thereby reducing the amount of contaminant flushed though or out of the soil or aquifer.

Phytoaccumulation: Phytoaccumulation is the uptake of contaminants by plant roots and the translocation/

accumulation (phytoextraction) of contaminants into plant shoots and leaves.

Phytodegradation: Phytodegradation is the metabolism of contaminants within plant tissues. Plants produce enzymes, such as dehalogenase and oxygenase, that help catalyse degradation. Investigations are proceeding to determine if both aromatic and chlorinated aliphatic compounds are amenable to phytodegradation.

Phytostabilisation: Phytostabilisation is the phenomenon of production of chemical compounds by plant to immobilise contaminants at the interface of roots and soil.

Chemical Oxidation

The Chemical oxidants most commonly employed to date include peroxide, ozone, and permanganate. These oxidants have been able to cause the rapid and complete chemical destruction of many toxic organic chemicals; other organics are amenable to partial degradation as an aid to subsequent bioremediation. In general the oxidants have been capable of achieving high treatment efficiencies (*e.g.*, > 90 per cent) for unsaturated aliphatic (*e.g.*, trichloroethylene [TCE]) and aromatic compounds (*e.g.*, benzene), with very fast reaction rates (90 per cent destruction in minutes). Field applications have clearly affirmed that matching the oxidant and *in situ* delivery system to the contaminants of concern (COCs) and the site conditions is the key to successful implementation and achieving performance goals.

Ozone Addition

Ozone gas can oxidise contaminants directly or through the formation of hydroxyl radicals. Like peroxide, ozone reactions are most effective in systems with acidic pH. The oxidation reaction proceeds with extremely fast, pseudo first order kinetics. Due to ozone's high reactivity and instability, O_3 is produced on-site, and it requires closely spaced delivery points (*e.g.*, air sparging wells). *In situ* decomposition of the ozone can lead to beneficial oxygenation and biostimulation.

Peroxide

Oxidation using liquid hydrogen peroxide (H_2O_2) in the presence of native or supplemental ferrous iron (Fe^{+2}) produces Fenton's Reagent which yields free hydroxyl radicals (OH^-). These strong, non-specific oxidants can rapidly degrade a variety of organic compounds. Fenton's Reagent oxidation is most effective under very acidic pH (e.g., pH 2 to 4) and becomes ineffective under moderate to strongly alkaline conditions. The reactions are extremely rapid and follow second-order kinetics.

Permanganate: The reaction stoichiometry of permanganate (typically provided as liquid or solid $KMnO_4$, but also available in Na, Ca, or Mg salts) in natural systems is complex. Due to its multiple valence states and mineral forms, Mn can participate in numerous reactions. The reactions proceed at a somewhat slower rate than the previous two reactions, according to second order kinetics. Depending on pH, the reaction can include destruction by direct electron transfer or free radical advanced oxidation—permanganate reactions are effective over a pH range of 3.5 to 12.

Applicability

The rate and extent of degradation of a target COC are dictated by the properties of the chemical itself and its susceptibility to oxidative degradation as well as the matrix conditions, most notably, pH, temperature, the concentration of oxidant, and the concentration of other oxidant-consuming substances such as natural organic matter and reduced minerals as well as carbonate and other free radical scavengers. Given the relatively indiscriminate and rapid rate of reaction of the oxidants with reduced substances, the method of delivery and distribution throughout a subsurface region is of paramount importance. Oxidant delivery systems often employ vertical or horizontal injection wells and sparge points with forced advection to rapidly move the oxidant into the subsurface.

Permanganate is relatively more stable and relatively more

persistent in the subsurface; as a result, it can migrate by diffusive processes. Consideration also must be given to the effects of oxidation on the system. All three oxidation reactions can decrease the pH if the system is not buffered effectively. Other potential oxidation-induced effects include: colloid genesis leading to reduced permeability; mobilisation of redox-sensitive and exchangeable sorbed metals; possible formation of toxic by-products; evolution of heat and gas; and biological perturbation

Limitations

The following factors may limit the applicability and effectiveness of chemcial oxidation include:

- Requirement for handling large quantities of hazardous oxidising chemicals due to the oxidant demand of the target organic chemicals and the unproductive oxidant consumption of the formation.
- Some COCs are resistant to oxidation.
- There is a potential for process-induced detrimental effects. Further research and development is ongoing to advance the science and engineering of *in situ* chemical oxidation and to increase its overall cost effectivenes.

Ex Situ Biological Treatment

Contaminants in extracted ground water are put into contact with microorganisms in attached or suspended growth biological reactors. In suspended systems, such as activated sludge, contaminated ground water is circulated in an aeration basin. In attached systems, such as rotating biological contractors and trickling filters, microorganisms are established on an inert support matrix.

Bioreactors degrade contaminants in water with microorganisms through attached or suspended biological systems. In suspended growth systems, such as activated sludge, fluidised beds, or sequencing batch reactors, contaminated ground water is circulated in an aeration basin where a microbial population aerobically degrades organic matter and produces CO_2, H_2O, and new cells. The cells form a sludge, which is settled out in a clarifier, and is either recycled to the aeration basin or disposed. In attached growth systems, such as upflow fixed film bioreactors, rotating biological contactors (RBCs), and trickling filters, microorganisms are established on an inert support matrix to aerobically degrade water contaminants.

One promising methodology includes the use of active supports (such as activated carbon, which adsorbs the contaminant and slowly releases it to the microorganisms for degradation). The microbial population may be derived either from the contaminant source or from an inoculum of organisms specific to a contaminant. Other applications include wetland ecosystems and column reactors.

Nutrients are often added to the bioreactors to support the growth of microorganisms. Bioreactors are a long-term technology. The process may take up to several years.

Trickling Filter: Another aerating wastewater treatment is the trickling filter. The trickling filter consists of a bed of highly permeable media, a water distributor, and an underdrain system. Wastewater is distributed over the top of the filter bed through which wastewater is trickled. The organic contaminants in wastewater are degraded by the microorganisms attached to the filter medium. The filter media may be rocks, plastic, or wood. The filter bed is normally round with depth varying from 3 to 8 ft (0.9 to 0.5m) and average 6 ft (1.8m). As wastewater flows over the solid filter media, it is aerated and the organic contaminants are degraded by the microorganisms attached to the media surface. The underdrain system is used to collect the treated water and any biomass detached from the filter media. It is also important as a porous structure through which air can circulate.

The duration of operation and maintenance of sprinkler

irrigation depends on the amount of time needed to capture and treat the contaminated waste; monitoring of treated water; and monitoring of potential metal accumulation.

Bioreactors are used primarily to treat SVOCs, fuel hydrocarbons, and any biodegradable organic material. The process may be less effective for some pesticides. Successful pilot-scale field studies have been conducted on some halogenated compounds, such as PCP, chlorobenzene and dichloro-benzene isomers. Bioreactors with cometabolites are used to treat PCBs, halogenated VOCs, and SVOCs in extracted ground water.

Limitations

The following factors may limit the applicability and effectiveness of the process:

- The dilute nature of contaminated ground water often will not support an adequate microbial population density, and this is especial true for suspended growth reactors. Nutrient addition may be necessary.
- Very high contaminant concentrations may be toxic to microorganisms and require special design approaches.
- Air pollution controls may need to be applied if there is volatilisation from activated sludge processes.
- Low ambient temperatures significantly decrease biodegradation rates, resulting in longer cleanup times or increased costs for heating.
- Nuisance microorganisms may preferentially colonise bioreactors, leading to reduced effectiveness.
- Residuals from sludge processes require treatment or disposal.
- Discharge of treated effluents may still be regulated.

Constructed Wetlands

The constructed wetlands-based treatment technology uses natural geochemical and biological processes inherent in an artificial wetland ecosystem to accumulate and remove metals, explosives, and other contaminants from influent waters. The process can use a filtration or degradation process.

Although the technology incorporates principal components of wetland ecosystems; including organic soils, microbial fauna, algae, and vascular plants; microbial activity is responsible for most of the remediation.

Influent waters with high metal concentrations and low pH flow through the aerobic and anaerobic zones of the wetland ecosystem. Metals are removed through ion exchange, adsorption, absorption, and precipitation with geochemical and microbial oxidation and reduction. Ion exchange occurs as metals in the water contact humic or other organic substances in the wetland. Wetlands constructed for this purpose often have little or no soil instead they have straw, manure or compost. Oxidation and reduction reactions catalysed by bacteria that occur in the aerobic and anaerobic zones, respectively, play a major role in precipitating metals as hydroxides and sulphides. Precipitated and adsorbed metals settle in quiescent ponds or are filtered out as water percolates through the medium or the plants.

Influent water with explosive residues or other contaminants flows through and beneath the gravel surface of a gravel-based wetland. The wetland, using emergent plants, is a coupled anaerobic-aerobic system. The anaerobic cell uses plants in concert with natural microbes to degrade the contaminant. The aerobic, also known as the reciprocating cell, further improves water quality through continued exposure to the plants and the movement of water between cell compartments. Wetland treatment is a long-term technology intended to operate continously for years.

Constructed wetlands have most commonly been used in wastewater treatment for controlling organic matter; nutrients, such as nitrogen and phosphorus; and suspended sediments. The wetlands process is also suitable for controlling trace metals, and other toxic materials. Additionally, the treatment has been used to treat acid mine drainage generated by metal or coal mining activities. These wastes typically contain high metals concentrations and are acidic. The process can be adapted to treat neutral and basic tailings solutions.

Limitations

The wetlands remediation technology must be adjusted to account for differences in geology, terrain, trace metal composition, and climate in the metal mining regions of the western United States.

The following factors may limit the applicability and effectiveness of the process:

- The long-term effectiveness of constructed wetlands is not well known. Wetland ageing may be a problem which may contribute to a decrease in contaminant removal rates over time.
- The cost of building an artificial wetland varies considerably from project and may not be financially viable for many sites.
- Temperature and fluctuations in flow affect wetland function and can cause a wetland to display inconsistent contaminant removal rates.
- Colder conditions slow the rate at which the wetland is able break down contaminants.
- A heavy flow of incoming water can overload the removal mechanisms in a wetland, while a dry spell can damage plants and severely limit wetland function.

2

Role of Pesticides

Biodegradation and Bioremediation

The rapidly growing population along with a move towards urbanisation and industrialisation has resulted in the accumulation of a wide variety of chemicals, dyes, fertilizers, pesticides and explosives, etc. in our environment. As a result of this chemical boom, many hazardous compounds are being added on a regular basis in our biosphere, the list being endless. Some of these compounds get degraded by natural means, although the rate of their degradation may vary according to the complexity of the compound. However, among them are some compounds, which persist for years and are resistant to normal mechanisms of degradation, these being categorised under 'Persistent organic pollutants' (POPs). The continuous influx of these chemicals and their tendency to persist has endangered all forms of life on earth as we are confronted by the growing problem of soil, sediment, groundwater and atmospheric pollution.

Chemical pesticides constitute an important class of POPs and have become a necessary evil and a mainstay of our agricultural system. A pesticide is any substance or mixture of substances intended for preventing, destroying, repelling, or mitigating any pest. Pests can be insects, mice and other animals, unwanted plants (weeds), fungi, or microorganisms like bacteria and viruses. Though often misunderstood to refer only to insecticides, the term 'pesticide' encompasses a great range of diverse substance, falling into a number of broad groupings such as herbicides, insecticides, molluscicides and fungicides.

There are absolutely no doubts that pesticides have revolutionised agriculture by increasing the productivity and the quality and quantity of food, fibre and feed. Apart *from* agriculture, pesticides find use in several other sectors like:

(i) The industrial, commercial, municipal and military for rodent and weed/brush control around plant sites, etc.

(ii) Medical sector for control of mosquitoes and rodents, etc., and

(iii) Urban environments for termite control around structures and pest control in gardens.

Studies have also shown that pesticides can cause health problems such as birth defects, nerve damage, cancer, and other effects that might occur over a long period of time. World Health Organisation (WHO) has estimated one million pesticides poisoning cases and *20,000* deaths every year globally. However, the use of chemical pesticides cannot be discontinued despite their persistence in the environment, their tendency towards bioaccumulation and toxicity to non-target organisms, including human beings as well as the toxicological and environmental problems of its residues (Singh *et al.*, 1999).

Pesticide Usage in India

India, unlike many other developed countries, produces and uses many deadly pesticides that have been banned years back in other countries. Recovering from the euphoria of the Green revolution, India is now battling the residual effects of the extensively used chemical fertilizers and pesticides in the country's soil. The injudicious and indiscriminate use of these pesticides has resulted in widespread contamination of soil, drinking water and food products. The decade from 1980 to 1990 in India alone saw the area under pesticides in India increase a whopping 20-fold, from six million hectares to 125 million hectares. After a high annual consumption of nearly 75000 MT reached in the early 90's, interventions in the form of Integrated Pest Management (IPM) practices have only now started to show a declining trend in the use of pesticides in India. However, India's consumption of pesticides per hectare is low when compared with world averages-0.5 kg/ha against Korea's 6.60 kg/ha and Japan's 12.0 kg/ha. Despite this low consumption of pesticides, the contamination of food products in the country is alarming.

About 20 per cent of Indian food products contain pesticide residues above tolerance level compared to only 2 per cent globally. A nationwide survey of food contaminants, conducted by the Indian Council of Medical Research (ICMR), revealed excessive levels of a cancer-causing fungal product in cereals and toxic metals in other food products of India (Pesticide news, 1993). It has also detected high levels of pesticide residues in cows milk and metals like arsenic, cadmium and lead in infant canned products and in turmeric.

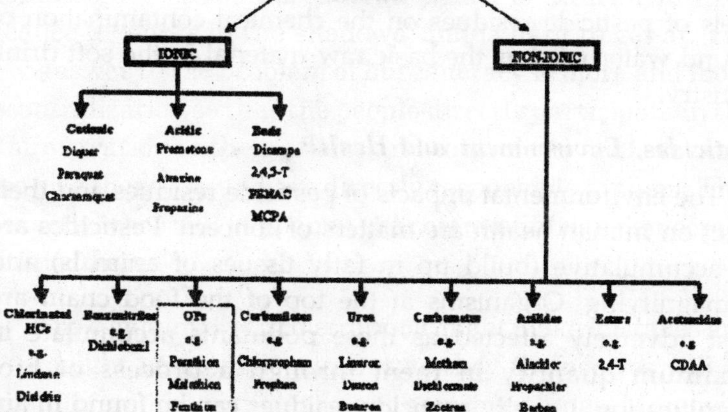

Figure: Classification of pesticides

Table: Consumption levels of various pesticides globally and in India

Types of pesticides	*Consumption*	*percentage (%)*
	Global	India
Insecticides	~29	~63
Herbicides	~4	~14
Fungicides	~21	~21
Others	~6	~2

A study by the Uttar Pradesh state's water work department in September 2003 (Environment News Network, 2003) reported that ground water sources in 36 districts in northern U.P. state were contaminated with traces of cadmium, fluoride, arsenic, nitrates and lead. Low concentrations of arsenic can slowly build up in the body, eventually causing cancers, skin diseases, and other illnesses. Prolonged exposure to arsenic can cause kidney, liver, intestinal, neurological, cardiovascular, and respiratory disorders. Recently, toxic pesticides in concentrations far higher than permitted by the European Economic Commission were found in 12 cold drink brands sold in and around New Delhi. Lab tests revealed that all samples, including those of Coke and Pepsi contained residues of four toxic pesticides and insecticides—lindane, DDT, malathion and chlorpyriphos in the soft drinks. The environmental organisation blamed the high levels of pesticide residues on the chemical contamination of ground water used as the basic raw material in the soft drink industry.

Pesticides, Environment and Health

The environmental impacts of pesticide residues and their effect on human health are matters of concern. Pesticides are bio-accumulative (build up in fatty tissues of animals) and biomagnifying. Organisms at the top of the food chain are most adversely affected as these pollutants accumulate in maximum quantity in them through a process of bio-magnification. Usually pesticide residues can be found in air, water and soil beyond levels considered as safe. Pesticides are known for causing deaths of other organisms like rare species of eagles, peacocks, fish and other aquatic life, bees, beneficial organisms in farming, etc. Extensive studies have shown that pesticides affect the soil health by killing the vital microbial fauna present in the soil. Each gram of soil may contain millions of microbes that are important for sustaining plant life. Pesticides not only kill the microbes in soil, they persist in the environment and accumulate in the food chain. Pesticides are carcinogenic, mutagenic and teratogenic and are also reported to possess estrogenic activity (Malley, 1997; Ohtsubo *et al.*, 2003 and Piper, 2005).

Studies have shown that pesticides can cause health problems, such as birth defects, nerve damage, cancer and other effects that might occur over a long period of time. WHO estimates one million pesticide-poisoning cases and 20,000 deaths every year globally. Persistent organic pollutants (POPs) due to their lipophilic nature, have a tendency to accumulate in fatty tissue. Recent studies on the estrogen-like behaviour of DDTs and PCBs in humans suggest the implications of these compounds in breast cancer (Calle *et al.*, 2002).

Organophosphorous Pesticides

The agricultural industry has been using chemical pesticides since the early 1900's. During the Second World War, German chemists developed a new class of pesticides, organophosphorous (OP) compounds. Since then, more than thousand OP compounds have been synthesised; some being stockpiled as potent chemical warfare agents while some others have been recognised to be useful OP pesticides.

Organophosphates comprise a large family of compounds that have appeared in the environment in recent evolutionary times. These chemicals are applied to crops, buildings, ornamental plants and lawns. Agricultural uses include field applications on corn, cotton, canola, alfalfa, and nuts. The organophosphate pesticides (OPs) are among the most widely used classes of pesticides in the world and about 30 per cent of total pesticides being used in India belong to this category. Approximately 3.7 million pounds of OPs are applied to agricultural fields in Canada (Canadian water quality guidelines, 1996). Over 40,000 tons of OP pesticides are applied in the United States annually (Shimazu *et al.*, 2001) with application of about 75 million pounds (Bravo *et al.*, 2002). Exterminators use organophosphate pesticides in residential and commercial structures, and certain pest control products for cats and dogs also contain organophosphorus compounds. People are commonly exposed to organophosphate pesticides through eating fresh and processed vegetables, contacting pesticide-contaminated surfaces, breathing air near pesticide applications (both indoors and outdoors), and drinking pesticide-contaminated water.

The multiple uses and ubiquitous nature of these pesticides results in widespread exposure to these chemicals for most people. Parathion and methyl parathion are two popular OPs that are widely used. These hydrolyse to form *p*-nitrophenol (PNP), which subsequently builds up in the soil. The United States Environmental Protection Agency (USEPA) has listed PNP as a priority pollutant necessitating the removal of this compound from soil. The primary biochemical effect associated with high-level exposure to OP insecticides is inhibition of acetylcholinesterase (AChE) resulting in acetylcholine accumulation (Hertel, 1993; Shimazu *et al.*, 2001 and Garcia *et al.*, 2003). Loss of AChE leads to a variety of effects resulting from excessive nervous stimulation culminating in respiratory failure and death. Symptoms of exposure include nausea, headaches, twitching, trembling, excessive salivation and tearing, inability to breathe because of paralysis of the diaphragm and convulsions, etc.

The organophosphate pesticides are also highly toxic to freshwater fish, aquatic life, marine organisms and birds. The USEPA considered that the general agricultural uses of chlorpyrifos posed a serious hazard to wildlife and honeybees and in 1988 concluded that the use of diazinon in turf and golf course treatment posed a widespread and continuous hazard to birds because of the bird die off events linked to diazinon use (Bailey *et al.*, 2000 and Hunt *et al.*, 2003). Residential uses of chlorpyrifos and diazinon have been banned by the USEPA. In a study in India, it was found that through breast milk, infants consumed 4.1 times more malathion than the average daily intake levels recommended by the World Health Organisation.

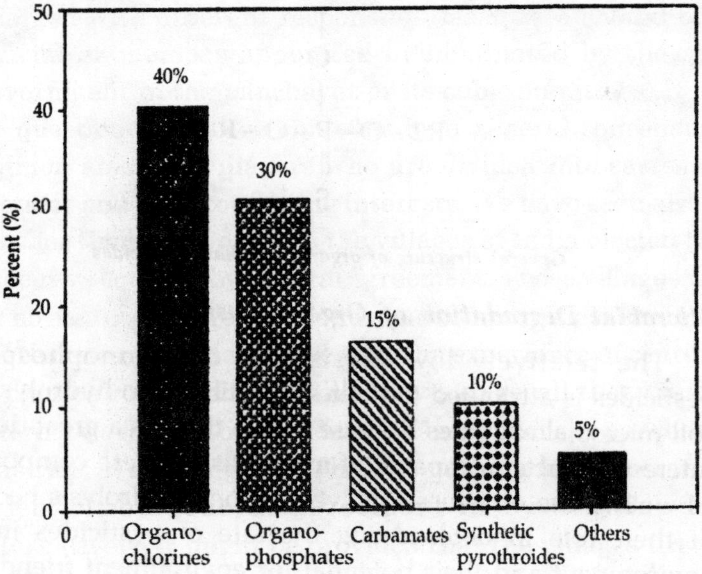

A correlation analysis between mothers' age and the content of the chemicals accumulated in breast milk indicated a substantial degree of correlation for malathion (Sanghi *et al.*, 2003). In an analysis of pesticide residues in soft drinks in India, chlorpyriphos was present in 100 per cent of the samples and in 75 per cent of the samples, chlorpyriphos residues exceeded the recommended limit by 3.9 to 7.6 times. In spite of the useful properties of organophosphates such as low persistence and high potency at low doses, their widespread and injudicious usage has caused environmental and health problems.

Chemical Structure

Organophosphate pesticides are sometimes called phosphotriesters as they contain three phosphoester linkages in which a central phosphorus atom is linked either to an oxygen or a sulphur; in addition they have R1 and R2 groups that are either ethyl or methyl in structure, and a leaving group R which is specific to the individual organophosphate.

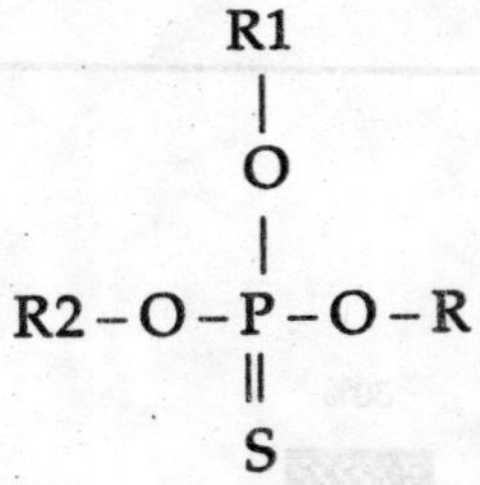

General structure of organophosphate pesticides

Microbial Degradation of Organophosphates

The relatively low persistence of organophosphate pesticides is attributed to their susceptibility to hydrolysis by soil microbial enzymes Because of this there is a great deal of interest in bacteria capable of metabolising these compounds as well as the enzymes involved in their hydrolysis because of their role in determining the fate of pesticides in the environment and their potential for environment friendly in situ detoxification/bioremediation (Mulbry and Karns, 1989). Although, organophosphate pesticides are toxic to higher organisms, being powerful inhibitors of the enzyme acetylcholinesterase (De Frank, 1991) they provide an abundant source of nutrition for bacteria. Studies on microbial degradation are useful in the development of strategies for the detoxification of insecticides by microorganisms.

The principal reactions involved in the breakdown of organophosphates are hydrolysis, oxidation, alkylation and dealkylation (Singh *et al.*, 1999). Microbial degradation through hydrolysis of P-O-alkyl and P-O-aryl bonds as shown below is considered to be the most significant step in the degradation of organophosphate compounds.

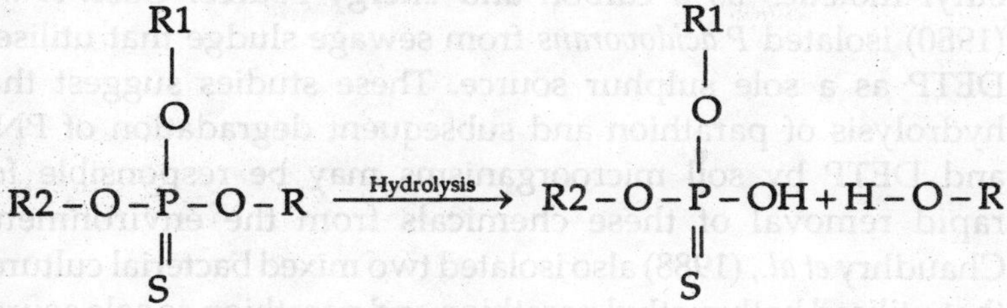

Parathion

Parathion (O,O-diethyl-O-4-nitrophenyl phosphorothioate) is the most well studied organophosphate pesticide in terms of its metabolism. Although it is relatively less persistent, this pesticide and its metabolites, particularly *p*-nitrophenol (PNP) have caused environmental problems. It is well established that microbial metabolism is the major means of parathion degradation in soil. The major degradation products have been identified as PNP and amino parathion (Liu *et al.*, 2005). The primary degradative pathway is initial hydrolysis to yield PNP and diethyl thiophosphoric acid (DETP). The secondary pathway involves the formation of paraoxon (diethyl *p*-nitrophenyl phosphate) that subsequently hydrolyses to form PNP and DETP. The third degradative pathway involves reduction of parathion under low oxygen conditions to yield *p*-aminoparathion followed by hydrolysis to *p*-aminophenol and DETP (Montgomery, 1997). Two *Pseudomonas* strains isolated from soil and sewage utilised parathion as source of carbon and hydrolysed it to PNP (Rosenberg and Alexander, 1979). A *Flavobacterium* sp. strain ATCC 27551 isolated from flooded soil hydrolysed both diethyl (parathion and diazonin) and dimethyl (methyl parathion and fenitrothion) phosphorothioates, while another organism *Pseudomonas* sp ATCC strain 29353, isolated from the same soil, hydrolysed only diethyl phosphorothioates (Adhya *et al.*, 1981). The second metabolite of parathion, DETP was found to be metabolised by enrichment culture obtained from cattle dip solution containing coumaphos (Cook *et al.*, 1980). These cultures mineralised DETP to sulphate and phosphate while utilising ethyl moieties as a carbon and energy source. Cook *et al.*, (1980) isolated *P.acidovorans* from sewage sludge that utilised DETP as a sole sulphur source. These studies suggest that hydrolysis of parathion and subsequent degradation of PNP and DETP by soil microorganisms may be responsible for rapid removal of these chemicals from the environment. Chaudhry *et al.*, (1988) also isolated two mixed bacterial cultures that utilised both methyl parathion and parathion as sole source of carbon.

A *Pseudomonas* sp. isolated from these mixed cultures degraded parathion to PNP only, while mixed cultures degraded the pesticides completely. The hydrolysis of parathion to DETP and PNP reduces the toxicity of parathion by more than 120 fold (Liu *et al.*, 2005).

Table: Microorganisms involved in the metabolism of parathion

Microorganism	*Metabolite*	*Reference*
Bacillus subtilis	Aminoparathion	Miyamoto *et al.*, 1966.
B. subtilis, B. megaterium, B. cereus, B. macerans, P. aeruginosa, P. ovalis, P. aurofaciens, Alcaligenes viscolactis, E. coli, Serratia plumuthica	Aminoparathion	Hirakoso, 1968.
Rhizobium japonicum, Rhizobium meliloti	Aminoparathion (major) PNP (minor)	Mick & Dahm, 1970.
Flavobacterium sp.	PNP	Sethunathan, & Yoshida, 1973.
Bacillus sp.	PNP	Siddaramappa *et al.*, 1973.
Mixed microbial cultures	PNP, Diethylthiophosphate, paraoxon and aminoparathion	Munnecke *et al.*, 1982
Pseudomonas stutzeri	Diethylthiophosphate and PNP	Daughton & Hsieh, 1977

The enzyme responsible for catalysing this hydrolytic step is organophosphate hydrolase (OPH E.C. 8.1.3.1). This enzyme has broad substrate specificity and catalyses the hydrolysis of several other organophosphate pesticides such as methyl parathion, paraoxon, dursban, coumaphos, cyanophos and diazonin and even some organophosphate chemical warfare agents such as sarin and soman (Richins *et al.*, 1997). The OPH enzyme is encoded by the *opd* gene. The *opd* gene has been isolated from taxonomically different bacteria and from geographically different regions of the world but the most widely characterised parathion degraders are *Flavobacterium* sp. ATCC 27551 isolated from a rice field in the Philippines (Sethunathan and Yoshida, 1973) and *Pseudomonas diminuta* MG isolated from Texas from an enrichment culture (Serder *et al.*, 1982) which possess high levels of this constitutively expressed enzyme. The *opd* gene has been cloned and sequenced from both *P. diminuta* MG (Serder *et al.*, 1989) and *Flavobacterium* ATCC 27551 (Mulbry, and Karns, 1989) and the genes were found to be identical although the plasmids carrying them were very different.

This suggests that the genes may be part of a mobile genetic element or transposon (Mulbry *et al.*, 1987). The widespread distribution of *opd* gene has been suggested to be due to its lateral transfer by a combination of transposition and plasmid transfer (Siddavattam *et al.*, 2003).

Considering the potential use of phosphotriesterases for the bioremediationof pesticides, substantial efforts have been made to understand the nature of the mechanism and substrate-binding determinants of OPH. OPH acts on a broad range of substrates, some of which are turned over at near diffusion rates whereas others are processed at more modest rates (Caldwell *et al.*, 1991; Scanlan and Reid, 1995). It belongs to a superfamily of metal-containing amidohydrolases (Holm and Sander, 1997). The crystal structure of OPH has been elucidated (Benning *et al.*, 2001) and it is reported to be a homodimeric protein with a monomeric molecular weight of 35 kDa (DiSoudi, *et al.*, 1999). Structures of OPH have been obtained with the inhibitors diethyl 4-methylbenzylphosphonate, diisopropyl methyl phosphonate and triethyl phosphate (Vanhooke *et al.*, 1996). These inhibitors are substrate analogues and bind in the active site of the enzyme giving indications of how the protein binds substrate. Pockets to accommodate the phosphate substituents have also been identified. Structural and kinetic studies carried out with this enzyme have led researchers to propose a plausible enzyme mechanism: it involves a single in-line displacement attack (S_N2) with the metal-bridging hydroxide ion acting as the nucleophile (Benning *et al.*, 2000).

A naturally occurring variant of OPH enzyme designated OPDA capable of degrading broader range of organophosphates was recently isolated from an *Agrobacterium radiobacter* strain (Horne *et al.*, 2002) which was very similar in sequence to the OPH from *P. diminuta* MG and *Flavobacterium* sp. ATCC 27551. The sequences of OPDA and OPH were found to be similar with 90 per cent identity at the amino acids level (Home *et al.*, 2002) and the sequence differences are thought to be responsible for the variation between the substrate specificities of OPH and OPDA. The latter has higher k_{cat} values for substrates with shorter side chains and can hydrolyse fenthion and phosmet for which OPH has no activity (Home *et al.*, 2002). Siddavattam *et al.*, (2003) recently described a novel gene in *Flavobacterium* sp. ATCC 27551 linked to *opd* that is involved in the degradation of PNP and is likely to act in concert with *opd* in the degradation of parathion.

Methyl Parathion

Methyl parathion (O, O-dimethyl O-(*p*-nitrophenyl) phosphorothioate) is widely used throughout the world and its residues are regularly detected in a range of fruits and vegetables. It is a highly toxic pesticide classified by WHO in the class of extremely hazardous pesticides and its high acute toxicity has prompted several countries like USA, Indonesia, Sri Lanka and Tanzania to ban its use, and several others like Colombia, Korea, China and Japan to restrict its use. However, it is regularly misused in many developing countries. Parathion and methyl parathion have similar chemical structures so it seems reasonable that the microorganisms and enzymes involved in the degradation of these two pesticides may also have homology with each other. Indeed, a methyl parathion degrading *Pseudomonas* sp. possessed DNA homologous to the *opd* gene encoding OPH (Chaudhry *et al.*, 1988). A P*seudomonas* sp. strain 50541 isolated from a pesticide waste disposal site was found to oxidise methyl parathion rapidly and completely (Chapalamagdu and Chaudhry, 1992). However, Zhongli *et al.*, (2001) reported a novel gene designated *mpd* from another methyl parathion degrading *Pseudomonas* sp. strain M6 that is completely different from *opd* genes. *Pseudomonas putida,* isolated from a rice field soil by Lakshmi and Lalithakumari, (1994), utilised methyl parathion as the sole source of carbon and/or phosphorus and could also utilise the intermediates of methyl parathion degradation, i.e. PNP, hydroquinone and 1,2,4 benzenetriol (the breakdown products of PNP).

The bacteria produced the enzyme organophosphorus acid anhydrase that hydrolysed methyl parathion to PNP (Lakshmi and Lalithakumari, 1994). A *Flavobacterium balustinum* strain has also been isolated from agricultural soils in Andhra Pradesh, India that could utilise methyl parathion as sole source of carbon. A large indigenous plasmid involved in the degradation of OP pesticides was identified in this soil isolate (Somara and Siddavattam, 1995). The plasmid borne *opd* gene of this organism was cloned and sequenced and the sequence was found to be 98 per cent similar to previously reported *opd* genes (Somara *et al.*, 2002). The major degradation product of methyl parathion is PNP as in case of parathion.

Fenitrothion

Fenitrothion (O,O-dimethyl O-(*p*-nitrotolyl) phosphorothioate) has been most widely and extensively used insecticide to control pests in agriculture and public health protection. Several pathways of fenitrothion exist in soil. Miyamoto *et al.*, (1966) studied the degradation of fenitrothion by *Bacillus subtilis.* The major metabolite was aminofenitrothion and minor metabolites were dimethyl thiophosphoric acid and desmethyl fenitrothion. A fenitrothion degrading bacterium *Burkholderia* sp. strain NF100 was isolated from fenitrothion treated soil.

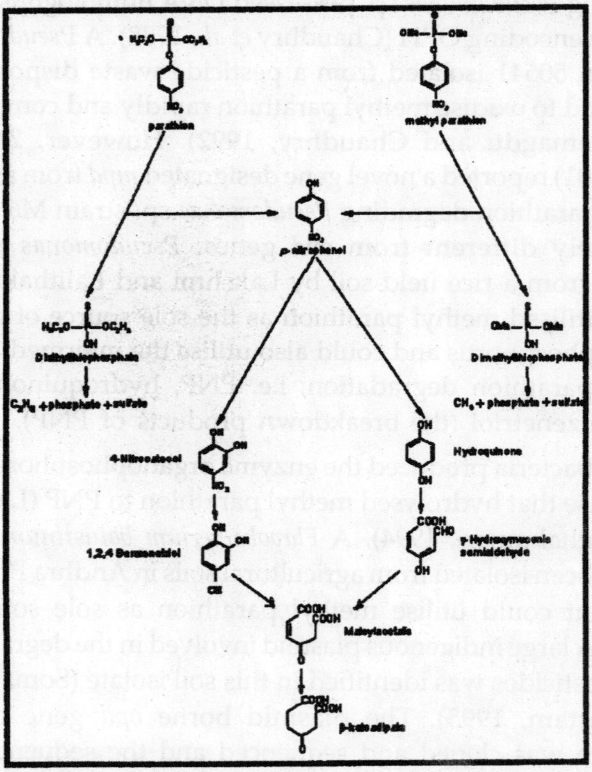

The degradation involved initial hydrolysis of phosphate bond to form 3-methyl-4-nitrophenol that was further metabolised to methylhydroquinone (Tago *et al.,* 2006).

Other Organophosphorous Pesticides

Diazinon, (0,0-diethyl-0-2-isopropyl-4-methyl-6-pyrimidye phosphorothioate) is one of the leading causes of acute insecticide poisoning for humans and wildlife. Not much is known about the microbial degradation of diazinon. Although *Flavobacterium* sp. ATCC 27551 was isolated originally as diazinon degrading bacterium, it was well characterised only with respect to parathion hydrolysis. Two *Pseudomonas* sp. isolated from sewage sludge degraded diazinon-producing DETP in the culture medium (Rosenberg and Alexander, 1979). A *Pseudomonas* sp. that hydrolysed several organophosphates including parathion, methyl parathion, dursban, paraoxon, aminoparathion and other methoxy or ethoxy substituted organophosphates, was found to hydrolyse diazinon as well (Bailey *et al.*, 2000). In another study, Barik *et al.*, (1982) found that cell-free extracts from two strains of *Arthrobacter* sp. SB3 and SB4, hydrolysed diazinon and the K_m values of the enzyme for the two strains were 1.3 and 2 imol/mg protein/min respectively. It had broad pH and temperature optima of 6-9 and 25-36°C respectively

Coumaphos (O,O-diethyl-O-(3-chloro-4-methyl-2-oxo-2H-l-benzo-pyran-7-yl) phosphorothioate) has a half life of about 300 days in soil and water (Mansee *et al.*, 2000). Three bacteria were isolated from enrichment cultures developed using cattle-dip vat solutions as inoculum. These isolates B1, B2 and B3 hydrolysed coumaphos to chlorferon and DETP. Chlorferon was further metabolised by Bl and B2 to a-chloro-b methyl-2, 3, 4-trigydroxy-trans-cinnamic acids (Shelton and Somich, 1988). Parathion hydrolase from *Flavobacterium* sp. 27551 was also found to hydrolyse coumaphos, yielding chlorferon and DETP (Manse *et al.*, 2005).

Attempts to isolate chlorpyrifos degrading bacteria from chlorpyrifos treated soils have generally been unsuccessful (Racke *et al.*, 1990, Mallick *et al.*, 1999) which has been attributed to the toxic effect of 3,5,6-trichloro-2-pyridinol (TCP), the main degradation product of chlorpyrifos on soil microorganisms (Racke *et al.*, 1990). Chlorpyrifos can nevertheless be degraded by bacteria capable of degrading other organophosphate pesticides such as *Flavobacterium* sp. ATCC 27551 and *Arthrobacter* sp. (Mallick *et al.*, 1999). However, Singh *et al.*, (2003) were able to isolate a bacterium capable of using chlorpyrifos as the sole carbon source.

p-Nitrophenol

Since the major metabolite of both parathion and methyl parathion is PNP, which has been listed as a priority pollutant by USEPA (Keith, and Telliard, 1979), its fate in the environment is of potential concern and much work has been done on this compound. PNP can be utilised as a source of carbon and energy by several bacterial species (Hanne *et al.*, 1993; Nishino and Spain, 1993; Leung *et al.*, 1997; Kadiyala and Spain, 1998 and Chauhan et. al., 2000). Two major pathways have been proposed for the aerobic bacterial degradation of PNP. In one pathway, which was elucidated in a *Moraxella* sp. by Spain and Gibson, (1991), PNP is degraded via the formation of hydroquinone followed by ring fission to form g-hydroxymuconic semi-aldehyde which is further converted to maleylacetate.

The other pathway which is more common in gram-negative bacteria, involves formation of 4-nitrocatechol followed by production of 1,2,4 benzenetriol which is then cleaved to maleylacetate (Jain *et al.*, 1994). Maleylacetate enters the TCA cycle via formation of b-ketoadipate and is thus mineralised to CO_2 and H_2O. In our laboratory, we have isolated three organisms *Arthrobacter protophormiae* RKJ100, *Burkholderia cepacia* RKJ200 and *Ralstonia* sp. SJ98 which can utilise PNP as the sole source of carbon and energy. We have elucidated the complete pathway of PNP degradation in strain RKJ100 that occurs via formation of benzoquinone, hydroquinone, maleylacetate and b-ketoadipate (Chauhan *et al.*, 2000). This pathway has been included in the University of Minnesota Biocatalysis/Biodegradation Database.

A plasmid of ~65 kb was found to be responsible for harbouring the genes for PNP degradation in this strain (Chauhan *et al.*, 2000). Kinetic studies on biodegradation of PNP by these strains have also been carried out (Bhushan *et al.*, 2000). Resting cells induced with PNP were used for studying the degradation kinetics for strains RKJ100 and SJ98. The apparent Vmax value of PNP degradation by strain RKJ100 was 0.28 mM and that for strain SJ98 was 0.23 mM as determined from Michaelis-Menten curves drawn between PNP concentration and rate of PNP degradation, while the maximum rates of PNP degradation (V values) were found to be 7.81 and 3.38 mM PNP degraded/min/mg dry biomass respectively, as determined from Lineweaver-Burk plots (Bhushan *et al.*, 2000). The intermediates of PNP degradation, benzoquinone and hydroquinone, were found to inhibit PNP degradation in strain RKJ100 non-competitively and competitively, respectively. In strain SJ98, these intermediates inhibited PNP degradation in an uncompetitive manner.

Very little work has been carried out on elucidating the molecular basis of *p*-nitrophenol degradation. Initial studies have suggested that genes encoding *p*-nitrophenol degradation are located on plasmids in case of *A. aurescens* TW17 (Hanne *et al.*, 1993) and *B. cepacia* RKJ200 (Prakash *et al.*, 1996). Bang and Zylstra, (1997) cloned a ~19 kb *Not*I fragment containing some of the gene(s) involved in *p*-nitrophenol degradation. In association with the hydroquinone pathway, the in *p*-nitrophenol degradation genes were cloned from *Pseudomonas* sp. strain ENV2030 and *Pseudomonas putida* JS444 (Zylstra *et al.*, 2000). In the case of the hydroxyquinol pathway, catabolic gene(s) for degradation of *p*-nitrophenol have not been reported. The only nucleotide sequence of in *p*-nitrophenol degradation genes *(nph*Al, *nph*A2*)* thus far reported is from *Rhodococcus* sp. strain PN1. The gene products NphA1 and NphA2 are reported to convert *p*-nitrophenol to 4-nitrocatechol, but the genes involved in the further degradation of 4-nitrocatechol and its metabolites from strain PN1 have not been reported to date (Takeo *et al.*, 2003). Therefore, more information about the catabolic genes involved in in *p*-nitrophenol metabolism is still needed, and in particular, information about those genes that act on the hydroxyquinol pathway is considered to be highly desirable. Kitagawa *et al.*, (2004) cloned three gene(s), namely *npc*A, *npc*B, *npc*C encoding some of enzyme(s) involved in p-nitrophenol degradation from *Rhodococcus opacus* SAO101. This gram-positive 4-NP degrader was originally isolated as a bacterium able to degrade dibenzo *p*- dioxin (Kimura and Urushigawa, 2001). The strain also degrades biphenyl, naphthalene, phenol, benzene, dibenzofuran, dibenzo-*p*-dioxin, and 4-NP. However, the molecular regulation of gene(s) involved in *p*-nitrophenol degradation has not yet been completely elucidated and the enzymes involved have not been purified and characterised. Therefore, this area of research still holds considerable challenge.

3-Methyl-4-nitrophenol

Although, there are several reports on biodegradation of PNP, little is known on the biodegradation of 3-methy-4-nitrophenol that is one of the major breakdown products of fenitrothion, a recalcitrant organophosphate insecticide. Being the non-polar methylated aromatic compound, 3-methyl-4-nitrophenol is highly toxic and, therefore, a complete degradation of this compound is important for environmental decontamination/bioremediation purposes.

Only recently, the involvement of a plasmid in the degradation of fenitrothion has been reported in which a *Burkholderia* sp. strain NF100 was shown to first hydrolyse the organophosphate bond of fenitrothion forming 3-methyl-4-nitrophenol that was further converted to methylhydroquinone, the substrate for oxygenase catalysed ring fission. The fenitrothion degradation ability was encoded on two plasmids pNFl and pNF2 (Hayatsu *et al.*, 2000).

Bhushan *et al.*, (2000) have reported the involvement of a gram negative, motile microorganism *Ralstonia* sp. SJ98 in the chemotaxis and biodegradation of 3- methy1-4-nitrophenol. The microorganism was capable of utilising both PNP and 3-methyl-4-nitrophenol as the sole source of carbon and energy. Catechol was found to be one of the major intermediate of the degradative pathway. Recently Tago *et al.*, (2006) have purified and biochemically characterised fenitrothion hydrolase, an enzyme involved in the degradation of fenitrothion from *Burkholderia* sp. NF1000. However, these pathways are yet not been fully characterised.

Genetic Engineering Studies

The slow growth and low culture yields of native OPH producing strains makes them uneconomical for practical use. Therefore, efforts have been made by several researchers to improve the applicability of OPH for pesticide bioremediation. The parathion hydrolase has been shown to be present on a 66-kb plasmid in *Pseudomonas diminuta* isolated from an enrichment culture (Serder and Gibson, 1985). The gene encoding this hydrolase has been cloned and introduced into various bacterial strains (Serdar and Gibson, 1985): *Streptomyces lividans* (Steiert *et al.*, 1989) and *E. coli* (Serder *et al.*, 1989), etc. Recombinant OPHs have been demonstrated to be equally effective in hydrolysing various organophosphorus pesticides (Phillips *et al.*, 1990; Xu *et al.*, 1996). Introduction of recombinant plasmids containing *opd* gene into *E. coli* under the control of its own promoter resulted in poor expression of the gene. However, when cloned downstream to an *E. coli* promoter sequence (such as the lac promoter), increased expression of the *opd* gene was achieved (Singh *et al.*, 1999). Both native and recombinant OPH have been used for the detoxification of organophosphate pesticides.

In a study by Richins *et al.*, (1997), the rate of hydrolysis of paraoxon and parathion was increased seven times by expressing OPHs on the cell surface of *E. coli* as a Lpp-OMP-OPH fusion. However, these recombinant strains were relatively unstable and hence a new plasmid was constructed to express OPH on the cell surface under control of a tightly regulated *tac* promoter. Various cell surface expression systems have been developed for gram- negative bacteria (Francisco *et al.*, 1992 and Fuchs *et al.*, 1991) Shimazu *et al.*, (2001) were able to genetically engineer a *Moraxella* sp. with functional expression of OPH on its surface which was capable of simultaneous degradation of parathion, methyl parathion, paraoxon as well as PNP. Walker and Keasling, (2002) engineered *P. putida* KT2442 with the genes responsible for parathion hydrolysis and mineralisation of PNP enabling this genetically engineered strain to use parathion as sole source of carbon and energy. Parathion formed a separate dense, non-aqueous liquid phase but was still bioavailable. In another study, directed evolution of OPH using a surface-display OPH library was used for screening of OPH variants with enhanced activity towards poorly degraded substrates. This approach resulted in a variant with methyl parathion hydrolysing activity that was 25 fold higher than that of the wild type (Cho *et al.*, 2002). Directed evolution was also used to evolve OPH into a series of mutant enzymes with catalytic properties similar to those of opdA. The mutants had improved catalytic efficiency for methyl-parathion, methyl-paraoxon and coumaphos-o-analogue (Yang *et al.*, 2003).

A consortium comprising two bacteria that were genetically engineered to each carry out part of the pathway for parathion degradation (*E. coli* carrying gene for parathion hydrolase and *P. putida* KT2440 containing a PNP-inducible plasmid borne operon encoding the genes for PNP mineralisation) effectively worked together to break down parathion and prevent accumulation of PNP (Gilbert *et al.*, 2003).

Application of Microorganisms for the Bioremediation of Organophosphorous Pesticides

Bioremediation is a technology that utilises the metabolic potential of microorganisms to clean up contaminated environments. Over billions of years, microorganisms have evolved an extensive range of enzymes, pathways and control mechanisms in order to degrade an array of compounds. In addition, microbes have the unique biochemical asset of being able to use molecular oxygen to catalyse the oxidation of numerous natural and man-made chemicals thereby initiating reaction sequences that enter the central pathway of metabolism such as Kreb's cycle or the fatty acid spiral. Microbial degradation critically affects the fate and behaviour of pesticides in soil. Ideally, bioremediation results in the complete mineralisation of contaminants to O_2 and CO_2 without the build up of intermediates. (Alexander, M. 1999 and Singh,

A. & Ward, O.P., 2004). Bioremediation processes can be broadly categorised into two types: *ex situ* and *in situ*. *Ex situ* bioremediation technologies include bioreactors, biofilters, land farming and some composting methods. *In situ* bioremediation technologies include bioaugmentation, bioventing, biosparging, biostimulation, liquid delivery systems and some composting methods. *In situ* treatments tend to be more attractive because they require less equipment, generally cost less and are more environment friendly (Crawford, R.L. and Crawford, D.L., 1998). One important characteristic of bioremediation is that it is carried out in non-sterile open environments that contain a variety of organisms. Of these, bacteria, such as those capable of degrading pollutants, usually have central roles in bioremediation, whereas other organisms (e.g. fungi and grazing protozoa) also affect the process. A deeper understanding of the microbial ecology of contaminated sites is therefore necessary to further improve bioremediation processes. The natural environment harbours a wide range of microorganisms that have crucial roles in bioremediation.

Bioremediation using white-rot fungi is a promising technology that is currently being researched extensively. This technology can be used in an *ex situ* or *in situ* manner. The white-rot fungi are a physiological rather than taxonomic grouping, comprising those fungi that are of extensively degrading lignin (a heterogenous polyphenolic polymer) within lignocellulosic substrates. Most studies focus on the ability of *Phanerochaete chrysosporium* to degrade persistent compounds, but *Phanerochaete sordida, Pleuotus ostreatus, Phellinus weirii,* and *Polyporus versicolor* have also been successful in laboratory studies (Safferman *et al.*, 1995). These fungi are capable of degrading chlordane, lindane and DDT, which makes them useful for the remediation of pesticide-contaminated sites (Alexander, 1999). The white rot fungus, *Phanerochaete chrysosporium* is capable of degrading DDT and also several chlorinated organic compounds (Morgan *et al.*, 1991; Paszezynski and Crawford, 1995). These fungi are also involved in the degradation of HCH. P. *chrysosporium* cultured under ligninolytic conditions has been reported to partially mineralise lindane in liquid cultures and in corn-cob amended silt loam soil. (Kennedy *et al.*, 1990). Fungal lignin degrading system (LDS) involved in degradation mainly consists of lignin peroxidase (LIP), manganese peroxidase (MnP), and laccase. The white rot fungi have been shown to produce different combinations of these extracellular enzymes that are involved in various mechanisms of transformation of pesticides. White-rot fungi could be used to inoculate a composing operation. However, large quantities of the fungus are required to remediate a site due to the very slow nature of compound degradation (Alexander, 1999).

Future Perspectives

The very high rate of population growth in developing countries still requires high yields of crops from existing agricultural land. Thus, for high yield of food grains the use of chemicals fertilizers and pesticides is expected to increase in near future, especially in the developing countries like India. Environmental pollution due to pesticides residues therefore will continue and strategies like Biodegradation/ Bioremediation will have to be followed.

Bioremediation is a young and exciting field that has broad application in biology and biotechnology. The great versatility of microorganisms offers a simpler, inexpensive and more environmentally friendly strategy to reduce environmental pollution than non-biological options. Recent advances in the identification of genes encoding pesticide degrading enzymes, their heterologous gene expression, library construction, vector design, and screening have improved this technology to a great extent. The current technology is sufficiently powerful for solving real world problems and discover new biotechnological applications. Although a lot of work has been reported on the biodegradation of parathion and methyl parathion but there are hardly any reports of biodegradation of persistent organophosphates such as chlorpyrifos, fenitrothion and quinalphos, etc. There are some reports on the *opd* genes, the picture of the mechanism of degradation is however not so clear. Thus, more research work still needs to be done so as to use the huge microbial potential for promoting the welfare and sustainable growth of our society. Recently, novel detoxification strategies using genetically engineered organisms with enhanced degradative capabilities have been studied. For example, the use of genetically engineered *E. coli* with surface expressed organophosphorus hydrolase, the enzyme responsible for drastically reducing the toxicity of organophosphate pesticides, for more efficient degradation of organophosphates (Richins *et al.*, 1997; Shimazu *et al.*, 2001 and Walker & Keasling, 2002). The use of immobilised enzymes for pesticide hydrolysis has also shown promising potential for pesticide detoxification (Richins *et al.* 1997). To enhance the overall degradation of less efficient substrates, enzymes with improved hydrolytic activity have also been developed with the help of techniques such as directed evolution (Yang *et al.*, 2003). Hence the development of biochemical and genetic techniques has expanded the potential for experimental advances in biodegradation so that it has become practical to isolate and modify useful genes from diverse microbial strains. The future holds great promise for continuing advances in basic research and for successful application of this research for solving the environmental problems posed by these chemicals.

3

Role of Probiotics

As the wild capture fishes are overexploited and on a declining state, shrimp and other fin fish culture is wide spread throughout the tropical area of the world. It is an industry set for a period of strongly growing demand, and is currently worth more than US $ 10 billion per year. Culture of fin fish, shell fish, algal plants are among the fastest growing food-producing sectors. This intensification of different aquaculture practices has led to the development of various environmental and disease problems. So far the conventional approaches are concerned, the use of disinfectants and anti-microbial drugs have had limited success in the prevention or cure of aquatic diseases (Subasinghe, 1997). Furthermore, the massive use of anti-microbials for disease control and growth promotion in aquatic sector encourages the natural emergence of resistant bacteria (WHO, anti-microbial resistance fact sheet 194).

So, much emphasis should be given on the reduced use of anti-microbials and chemicals in aquaculture industries.

Thus, how to improve the ecological environment of aquaculture has become the focus of attention of international level. Now, researchers are trying to use probiotic bacteria in aquaculture to improve water quality by balancing bacteria population in water and reducing bacterial population (Verschuere, 2000).

Probiotic is live beneficial bacteria that reduce the harmful microbial load and fish growth. Researchers are paying more attention to this new approach i.e "eco-friendly aquaculture", and have made considerable headway.

Concept of Probiotics

The term 'probiotics' is Greek word means 'for life', composed of two words; 'Pro'- favour, and 'Bios' - life, that means probiotic is a substance that favours life. Being an antonym of antibiotics, probiotics involves multiplying few good/useful microbes to compete with harmful ones, thus suppressing their growth. These are certain bacteria and yeast that promote life and are not harmful on continued used for a long time (Mishra *et al.*, 2001). The term 'probiotics' was first coined by Lilly and Stillwell (1965) who proposed the definition, "substances produced by one protozoan that stimulated the growth of another'. Parker (1992) defined it as, "organisms and substances which contribute to intestinal microbial balance". Fuller (1992) defined it precisely as, "live microbial feed supplement which beneficially affects the host animal by improving its intestinal balance". Ruiz *et al.* (1998) also suggested probiotics as, "beneficial microorganism which can protect organisms against pathogens or enhance their growth". Jory (1998) define probiotics from aquaculture point of view as, "culture (single or mixed) of selected strains of bacteria that are used in culture and production system (tanks, ponds, and others) to modify or manipulate the microbial communities in water and sediment, reduce or eliminate selected pathogenic species of microorganisms, and generally improve growth and survival of the species".

There are several classes of bacteria that can be used as potential probiotics in aquaculture. These are: lactic acid bacteria (LAB such as carnobacterium, Lactobacillus (Gram positive, rod) and Sterptococci (S. *faecium)* (gram negative, cocci). The also includes several groups like Vibrio *(V. alginolyticus)*, Bacillus (Bacillus IP5832), Pseudomonas, Bifidobacterium, Actinomyces. Several fungi, like Aspergillus, yeast, like Sacchromyces, microalgae *(Teraselmis suecica)* (Austin *et al.*, 1992) also have probiotic properties in them. In coastal ecosystem, Bacillus, Nitrosomonas, Nitrobacter are dominant as probiotic bacteria (Dalmin *et al.*, 2001).

Probiotics Development

Aquaculturists throughout the world have been using a wide array of chemicals and biological agents in order to improve soil and water quality and to thereby prevent or control different diseases.

Several groups of substances are being used in aquaculture practices, viz. fertilizers, oxidising agents, peroxides and chlorine compounds, antibiotics, etc.

Fertilizers are highly soluble and release nutrients to water and as such when applied more may cause eutrophication. Due to eutrophication, dense algal bloom may appear on water surface some of which release toxins and thereby create fish susceptible to disease. Nitrogen fertilizers when used in excess amount can result in toxic ammonia concentration in water. Ammonia fertilizers and urea may reduce pond pH because nitrification releases hydrogen ions. Manures can sometimes be contaminated by heavy metals (Rao, 2001). In some parts of world, aquaculturists use human waste as a fertilizer, which increases the risk of contamination of aquatic animals with potential human pathogens.

Oxidising agents like $KMnO_4$, peroxides chlorine compounds are used for killing disease causing organisms, oxidising organic and inorganic substances so that the Biochemical Oxygen Demand (BOD) and Chemical Oxygen Demand (COD) will be less. But, being toxic to phytoplankton, KMn_{O4} may reduce the production of Dissolved Oxygen (DO) through photosynthesis. Some organic constituents of water reacts with chlorine residuals to form toxic compounds like trihalomethances, dioxins and polychlorinated biphenyles (PCBs) which are persistant, carcinogenic, contaminate food and cause chronic infection to human (Rao, 2001).

Continued use of antibiotics may give rise to antibiotic-resistant strains that cause severe infection to fish and man also. Several strains of Salmonella, Aeromonas hydrophila, and Pseudomonas shigelloids developed resisitancy to several antibiotics like Nalidixic acid (11 %), Oxolinic acid (12%), Chloramphenicol (17%), Neomycin (6%), Oxyteracyclin (94%), Tetracyclin (75%), Furazolidaqne (47%), Sulphamethoxazole (25%) when these antibiotics were used in culture pond for long term basis (Twiddy and Reilly, 1995).

Resistant bacteria proliferate after an antibiotic has killed off the other bacteria. If antibiotic Kanamycin (50mg-1) was added to sea water, bacterial numbers decreased for about 2 days, but after 2 days, they recovered to the original level of 10^6 cells ml-1 (Maeda *et al.*, 1997). Resistant bacteria also can transfer their resistant genes to other bacteria (trans-ovarial/ horizontal transmission) that have never been exposed to the antibiotic.

Antibiotics are sometimes proved to be ineffective when resistant bacteria develop *(Vibrio vulnificus)* or the pathogens enter a particular phase or part of the environment where they are insensitive (Boonyarapalin, 1990; Nash, 1990). For these reasons, Norway has dropped antibiotic use approximately 50 metric tons/year in 1987 to 746.5 kg in 1997, but the production increased from 5×10^4 to 3.5×10^5 metric tons in the same time. This increase was due to use of effective vaccines (Lunestad, 1998; Subasinghe, 1997). However, for these inconveniences, researchers were engaged for finding some substances, which could eco-friendly, manage the aquatic environment. Yasuda and Taga (1980) anticipated that bacteria would be useful as food and as biocontrol agents of fish disease and activators of the rate of nutrient regeneration in aquaculture.

Use of Bacillus S11 as probiotic in Peneaus monodon

cultured resulted in siognificant differences (P<0.05) in growth and survival between probiotic treated control groups (Rengpipat *et al.*, 1998). Use of Vibrio alginolyticus as probiotic in many Ecuadoran shrimp hatcheries increased the production by 35 per cent. The result of such type of researches influenced the farmers to adopt this new technique which was first commercially utilised in shrimps/prawn production system Initially, probiotic research was focused on fish juveniles, recently more attention has been given to larvae of fish and shellfish and to live food organisms (Verschuere *et al.*, 2000).

Mode of Application

Based on their method of application probiotics can be classified in two groups.

Feed Probiotics

These are administered through feed to the fish/shellfish fish in the form of capsules paste, powder or granules. Another way is bioencapsulation. This is done by feeding the fishes/larvae with live food organisms like artemia, daphnia, tubifex, filamentous algae, etc. enriched with probiotics. Enrichment is carried out by rearing the organisms in media containing a high proportion of probiotic bacterial strains.

Effectiveness of this method depends upon the adherence of bacteria to these organisms. Generally, filter feeders contain more bacteria both on surface and in their gut. Bioencapsulation comes into use in hatchery rearing of larvae and also in aquarium fish management.

Water Probiotics

These are added directly into rearing media to modify the microbial composition of water and sediments. This can be performed in 3 ways:

(i) *Direct Cell Addition:* here microbial cells are cultured in liquid media which is centrifuged later and the aliquot is added into the water.

(ii) *Addition of Cell Suspension:* cells are harvested by centrifugation following the previous method, washed in normal saline solution (NSS- contains 0.85-1 per cent NaCl) and finally suspended in NSS and added into the culture medium.

(iii) *Broth Addition:* cells are allowed to grow to sufficient numbers in liquid medium and added directly into the water.

Functions and Mode of Action

The probiotic treatment may be considers as methods of biological control, the so-called "biocontrol" which indicates the limitation or the elimination of pests by the introduction of adverse organisms, like parasites or specific pathogens. 'Biocontrol' implies only that the strains are antagonistic to pathogens (Maeda *et al.*, 1997). When these microbes remain viable in the Gl (gastrointestinal) tract, then these are called 'probiotics' (Gatesoupe, 1999). Treatment of pollutants or waste by the use of microbes that break down the undesirable substances is called 'bioremediation'. The same concept is sometimes named 'bioaugmentation' (Moriarty, 1997, 1998).

The possible modes of action are as follows:

- Improvement of water quality.
- Production of inhibitory compounds.
- Competition for chemicals or available energy.
- Competition for adhesion sites.
- Enhancement for adhesion sites.
- Interaction with phytoplankton.
- Source of macro and micronutrients.
- Enzymatic contribution to digestion.

Improvement of Water Quality

In several studies, water quality has been recorded during the addition of the probiotics. But the success varies greatly, depending on the nature of the technical information available to the end users. The bacteria that are added must be selected for specific function that are amenable to bioremediation, and added at a high enough population density, and under the right environmental conditions, say for, Nitrosomonas, Nitrobacter can reduce ammonia and nitrite whereas Bacillus can efficiently degrade organic matter.

Their mechanisms of action range from the production of organic acids which lower pH, to production of hydrogen peroxides which can be toxic to pathogens, and metabolites which can neutralise the toxins released by pathogens. They also produce substance, which aid in the detoxification of injurious metabolites.

Gram-positive Bacillus sp. are generally more efficient in converting organic matter back to CO_2 than the gram-negative bacteria, which would convert a grater percentage of organic carbon to bacteria biomass or slime (Stanier *et al.*, 1963). Bacillus decomposes the organic waste by releasing ex'oenzymes. However, their natural being the bottom sediments, they normally present in low numbers in water column. Decomposition of organic matters in the bottom result in better penetration of oxygen into the sediment providing a better environment for culture with no room for the anaerobes which might result in disease outbreak (Kumar and Sharma, 2001).

No obvious an effect of Bacillus S11 on water quality was found (DO, Temperature, Salinity, Ammonium, Nitrite, Phosphate, etc.) when added to P. monodon culture tanks (Rengpipat *et al.*, 1998).

Aqua. Bact. Aid R which consisted of bacillus, Nitrobacter, Pseudomonas (two strains), Aerobacter, Cellulomonas, and Rhodopseudomonas, showed little promise for improving water quality, with the exception of slight improvement in DO concentrations when tested under laboratory condition (Boyd *et al.*, 1984).

They performed another experiment by treating 4 channel catfish ponds with a bacterial inoculum containing 7 species of bacteria where no significant difference was observed ($P>0.05$) in inorganic nitrogen concentrations, total phosphorus, COD, BOD_5 phytoplankton, per millilitre.

Eaton *et al.* (1995) used the bacterial inoculum in water and found an increase in COD in treated ponds (42.7 mg^{-1}) compared to that in control ponds (31.7 mg^{-1})

When Bacillus spp. was applied to channel catfish *Ictalurus punctatus* ponds at Auburn, Alabama, USA,3 times per week, few significant differences ($P<0.1$) in concentrations of water quality variables were observed between ponds treated with bacteria and control pond (Quiroz and Boyd, 1998). Reduction in organic matter and ammonia in both fresh and saline water on application of probiotic bacteria was reported by Douillet (1998). Nutrients such as nitrate, phosphate and silicate were also found to be more in P. *monodon* experimental tanks compared to controls on provisions of probiotics (Ravi *et al.*, 1998).

Naik *et al.* (1999), reported that total alkalinity was lowered with higher rate of G-probiotics [Stafac-20, contains 2 per cent iriginiamycin] applied.

Decrease in transparency (40-63 cm in control ponds and 24-54 cm in treated ponds) was reported by Prabhu *et al.* (1999). They applied a commercial blend, i.e., NS-Series Super SPO. PH values were more in experimental ponds (7.7-8.5) compared to controls (7.3-8.0). Ammonia level always declined after application of probiotics and remained below critical level of less than 1 ppm whereas in control ponds it exceeds 2 ppm in some cases. Perhaps, reduction in ammonia might be due to nitrification as result showed increased level of nitrate in experimental tanks, whereas control ponds showed continuous increase in ammonia levels. Phosphate and silicate value increased in experimental ponds whereas both were decreased in control ponds (from 0.0094mgl^{-1} to 0.0031 mgl^{-1} and from 0.0036 mgl^{-1} to 0.0020 mgl^{-1} respectively). This increment proved the increased mineralisation.

Increase in DO and pH values on application of a commercial probiotic were reported by Sharma and Bhukhar (2000). They used a commercial preparation 'Aquazyn which contains several bacteria, viz. *Bacillus subtillis, B. magaterium, B. licheniformis, B. polymyxa, Remenococcus albus, AspergiHus orygae.* Orthophosphate concentration was little influenced whereas nitrate-nitrogen concentration was remain unchanged. Hepher (1967) reported a decline in nitrate-nitrogen in alkaline waters.

Hirata *et al.* (2001), conducted such experiments on carps *(Cyprinus carpio)* with a commercial product 'Amakusa-PX' (98 per cent eelmash + 2 per cent probiotics). Result showed better primary production (85± 53x10^6 cells/ml), lower lower levels of ammonia (30 per cent lesser than control), and higher growth (30 per cent higher than control).

Criticism of the practice of probiotic use has been made by Dr. C.E. Boyd (1998). He found that the bacteria production he tried did nit work as claimed. Probably, this may be due to lack of sufficient amount of right strain bacteria in those production or the bacteria were not viable. Moreover, the physiological and ecological requirements of those bacteria may not be present in the culture medium (Moriarty, 1996).

In China, the studies on probiotics in aquaculture were focused on the photosynthetic bacteria. Qiao Zhenguo *et al.* (1992) reported that addition of photosynthetic bacteria in the food or culture water. By using Rhodomonas sp. (photosynthetic bacteria) Cui Jingjin *et al.* (1997) also reported similar observations.

Photosynthetic bacteria could eliminate the NH_3-N, H_2S and organic acids, and other harmful materials rapidly, improve the water quality, and balance the pH (LiZhuojia *et al.*, 1997). Heterotrophic probioticbacteria may have chemical actions such as oxidation, ammonification, nitrification, dentitrification, sulphurication and nitrogen fixation. Such mineralisation processes produced salts and nutrient that promote phytoplankton growth.

Biofilters

Accumulation of dissolved nitrogen, especially ammonium (NH_3+NH_4^+), as a result of food addition and excretion of organisms reared at high density, is one of the main problems in intensive culture systems (mainly shrimp), affecting their food ingestion, growth and survival rates (Tomasso, 1994; Wasielesky *et al.*, 1994; Ostrensky and Wasielesky, 1995; Canalli ety al., 1996) and when this effluent, without any prior treatment, released may cause environmental degradation and disease outbreak (Hopkins *et al.*, 1995). An alternative way to maintain high water quality is biological treatment (Wheaton, 1997) using biofilter where pre-colonised microorganisms absorb excess nutrients from the water.

When any substrate is immersed in water having high organic load, microalgae grow and utilise nitrite and nitrate produced by nitrifying bacteria colonised in that firm. The microalgae (diatoms, filamentous cyanobacteria) also provide the food source for cultured organisms especially filter feeders.

The problem with this type of biofilm is that they harbour pathogenic bacteria like *Vibrio harveyi* with cause disease (Karunasagar *et al.*, 1996). The formation of biofilm also takes long time that create pollution to water quality and cause significant economical losses (Hochheimer and Wheaton, 1991). One alternative addition to biofilm to overcome these problems is start-up of biofilm by incorporating probiotic bacteria which reduce the settlement time for film (Carmigiani and Bennett, 1977) and prevent it from being colonised by pathogenic bacteria (Austin and Day, 1990; Alabi *et al.*, 1999). A comparative study was carried out in shrimp pond using established biofilter, where 'starter' bacterial population used and new biofilter. Establishment of chemical stability in ponds with new biofilter look more than 3 months and resulted in lower biomass production (2.2 kgm-^{3}), lower survival (43 per cent and higher water exchange (18.4 per cent/day) as compared to established biofilter ponds (2.5 kg m^{-3} biomass, 62 per cent survival rate and 7.2 per cent/day water exchange). The seeding of starter culture into biofilter at 10 per cent decrease the start-up time of a new filter by 81 per cent.(Gross *et al.*, 2003).

Thus, microinoculated biofilm maintains good water quality, provides foods for Daphnia (Langis *et al.*, 1988), Nile Tilapia (Shrestha and Kund-Hansen, 1994) and carp (Rameesh *et al.*, 1999), reduce the necessity of water exchange and contribution to the increase in weight of cultured organisms (Thompson *et al.*, 1999).

The water filtered through biofilter matured with microorganisms is called microbially matured water which contributed to lesser disease outbreak and higher survival of culture organisms (Vadstein *et al.*, 1993; Skjermo *et al.*, 1997).

Competition for Iron

Siderophores are low molecular weight (<1,500), ferric ion-specific agents (Neilands, 1981) which can dissolve precipitated iron and make them available for growth. Thus, siderophore-producing beneficial bacteria can inhibit the growth of opportunistic bacteria by making oron unavailable for growth. The possible effectiveness of siderophore-producing probiotics can be illustrated by the study of Gatesoupe (1997), in which addition of the bacterial siderphore deferoxamine to live food (rotifers) increase the resistance of turbot larvae challenged with pathogenic Vibrio strain P.

Interactions with Plankton

Fukami (1999) demonstrated that many bacterial strains might have significant algicidal effect on many species of microalgae, particularly on red tide plankton. This finding would be advantageous when undesired algal species develop in the cultured pond. A bacteria suspension consisted of Bacillus, Nitrobacter, Pseudomonas, Enterobacter, Cellulomonas and Rhodopseudomonas sp. applied to channel catfish ponds decreased the proportion of cyanobacteria (Boyd *et al.*, 1984).

Positive effects of bacteria on cultured microalgae have

also been observed (Fukami *et al.*, 1997; Munro *et al.*, 1995; Rico-Mora *et al.*, 1998; Suminto and Hirayama, 1997). Zooplankton population showed a significant increase when the water was treated with a commercial probiont Aquazyn having a maximum population of 1510 Nos./It. (Sharma and Bhukahr, 2000). Kumar (1994) and Kumar and Sharma (1999) also observed similar effects.

Production of Inhibitory Compounds

Microbial populations may release chemical substances that have a bacteriocidal or bacteriostatic effect on other microbial populations, which can alter interpopulation (Fredrickson and Stephanopoulos, 1981; Lemos *et al.*, 1991; Pybus *et al.*, 1994). The presence of bacteria producing inhibitory substances in the intestine of the host, on its surface, or in its culture medium is thought to constitute a barrier against the proliferation of (opportunistic) pathogens.

In general, the anti-bacterial effect of bacteria is due to the following factors, either singly or in combination:

- Production of antibiotics (Williams and Vickers, 1986).
- Production of bacteriocins (Bruno and Montville, 1993; Pybus *et al.*, 1994; Vandenbergh, 1993).
- Production of siderophores, lysozymes, proteases, and/or hydrogen peroxides and alteration of pH values by the production of organic acids (Sugita *et al.*, 1996).
- Vandenbergh (1993) added the formation of ammonia and diacetyl to this list.

Ringo and Gatesoupe (1998) discussed about the inhibition of bad bacteria by LAB that produce bactericins and organic acid named lactic acid. Inhibitory substance produce by Carnobacterium sp. may reduce the growth of V. anguillarum and A. salmonicida (Joborn *et al.*, 12997).

Disease Control

Previously several anti-microbial drugs were used to contain infections in culture system. But, since the probiotic has been come into used in aquaculture sector, disease outbreak is also controlled by probiotics simultaneously with other functions of it.

Probiotics can control/prevent disease by its array of functions that are listed below:

- Maintains favourable and healthy environment by reducing the pollution level.
- Kill/reduce the pathogenic bacterial population in water by secreting compounds or compete with them for nutrient and adhesion site.
- Exclude the pathogenic bacteria from gut microbiota of cultured organisms.

Production of anti-microbial compounds against fish pathogenic bacteria in vitro, are reported for bacteria isolated from the flora of both about (Dopazo *et al.*, 1988; Westerdahl *et al.*, 1991) and larval (Bergh, 1995) marine fish.

During challenge rest of *P. monodon* with Vibrio sp. *(V. harveyi* D331), 100 per cent survival observed in treatment groups whereas it was only 26 per cent with control groups. *P.monodon* was fed with Bacillus SII and may produce some unknown by-products negatively affected V. *hareyi* D331 (Hastings and Nealson, 1981). Bacilli inhibit pathogenic Vibrios by competing for nutrients, and damaging the slime layers of Vibrios through secreting enzymes that degrade slimes (Moriarty, 1997). Austin *et al.* (1995) recently found that the probiont *Vibrio alginolyticus* applied to salmon could reduce disease caused by *Aeromonas salmonicida, V. anguillarum* and *V. ordalii.*

Smith and Davey (1993) report that fluorescent strain pseudomonad bacteria can competitively inhibit the growth of fish pathogen *A. salmonicida.* Here, inhibition was due to competition for free iron. Kameri *et al.*, (1987) documented that anti-viral microorganisms are present in seawater that suppresses the viral growth. Based on this concept, anti-viral bacteria are added to larval tank.

Maeda *et al.* (1997) conducted an experiment with the strain VKM-124, *Pseudoalteromonas udina* that has the vibriostatic activity. The strain shows repressive activity towards Sima-aji Neuro Necrosis Virus (SJNNV), Baculo-like viruses. If adult fish, which feed on artificial compound food (AFC), are to be protected from viruses, the bacterial culture liquid is mixed with the AFC at 20 per cent (V/V) of AFC.

Yoshimizu *et al.* (1997) reported in their experiment that two isolates of Vibrio sp. have anti-viral activity against Infections Haematopoetic Necrosis Virus (IHNV), Onchorhynchus Masou Virus (OMV). Vibrios reduced the numbers of plaque.

Growth Promotion

Growth promotion by probiotics is attributed through:

- Providing food such as plankton to fish by booting up plankton production through minerlisation.
- It acts as food benthic and planktonic organisms which are subsequently transferred to higner trophic levels (Thompson *et al.*, 1999).
- It maintains favourable ecosystem.
- It excludes the pathogenic microbes from intestinal microbiota and secrets enzymes to enhance digestion and some times synthesis vitamins like B complex, K, etc.
- It represents a food source for shrimp larvae due to higher nitrogen and potassium contents.
- Due to better digestion of nutrients, food uptake and conversion to flash increase that ultimately give rise to better growth and production.

Several commercial probiotics like Livol induced highest growth at a dosage of 1 per cent of the diet in common carp (Abraham, 1992; Shadakashri, 1993), at 0.5 per cent level in Catla (Gireesha, 1993) and Rohu (Maheshappa, 1993).

Crown-gro-fast induced maximum growth in Mahaseer at

a level of 1.5 per cent of the diet (Shivkumar, 1995). Stafac-20 incorporated in a fish meal diet and gave higher weight in Catla fry any Mahseer fingerling and Rohu fry at 60 ppm and 100 ppm level respectively.

Enhancement of the Immune Response

Fish larvae, shrimp and other invertebrates unlike adult fish are dependent primarily on non-spcific immune reponses for their resistance to infection (Soderhall and Cerenius, 1998). Many reports are reviewed recently by Sakai (1999) where bacterial compounds were used as immunostimulant in fish and shrimp. It has also been suggested that ingestion of bacteria and subsequent endocytosis in cod and herring larvae are involved simulation of the developing immune system (Olafsen, 1998).

Constraints

One problem with probiotic bacteria especially LAB, that they can not sustain in culture medium (water) and fish intestine for more than 7 days. So, repeated application of probiotics to maintain population to desired level is essential which makes probiotic treatment costly to the farmers.

One should be aware of the risk that repeated sub cultivation of isolated bacterial strains in laboratory condition could change properties of the strains, making them unsuitable for use as probionts in the field.

The scarcity of data about their effective implementation in practice makes the practical relevance of probiotics in aquaculture hazy as yet. Furthermore, there is still a lack of knowledge about the exact modes of action involved in probiotic effects.

Another problem is probiotic application based on single, strain bacteria is less effective. So, in most commercial preparation, multiple strains are used to obtain/utilise their combined effect. Proper field research is needed about 'total probiotics' so that it can be commercialised more, to be applied in aquaculture sector more effectively.

Probiotic stimulate digestion and also depress post-metabolic activities that lead to the production of methane and ammonia which are toxic to fish (Naik *et al.*, 1999).

Conclusion

The feasibility and future of the application of probiotics in aquaculture has tremendous scope. At present, the probiotics are widely applied in United State of America, Japan, European countries, Indonesia and Thailand, with commendable results.

With the advent of aquaculture practices and its intensification, the scope of probiotic application is also increase. For water quality and nutrient maintenance, we may use several manures, fertilizers. For disease control, we may use several aqua-drugs. But probiotics may function simultaneously both. This new practice is eco-friendly also. Understanding the economic value and potential social benefits of the application of probiotics in aquaculture, recently government has paid more attention to the study and development of probiotics in aquaculture.

4

Periphyton based Recirculating System

Biotechnological Tool

Faced with the ever increasing population and the subsequent demand for food, aquaculture is gearing up for more production. While output from capture fisheries grew at annual average rate of 1.2 per cent, output from aquaculture (excluding aquatic plants) grew at a rate of 9.1 per cent. The latter is a faster rate than any other animal food producing systems such as fisheries and terrestrial farmed meat (FAO, 2003). The need to increase aquaculture production is driving the industry towards more intensive practices. Some of the factors that influence this trend are: limitations in quality and quantity of water, availability and cost of land, limitations on water discharges and environmental impacts. Recirculating technologies help minimise these issues (Gutierrez-Wing and Malone 2006). When the resources are getting reduced mainly due to competition for water and land, the only option to meet ever increasing demand for fish is the vertical expansion in terms of obtaining higher yields for a given critical resource.

One such critical resource could be water. As the water resources in many parts of the world are getting scarce day by day, it is pragmatic for the aquaculture industry to evolve systems which will reduce the usage of water and at the same time obtaining enhanced productivity in terms of mass produced per volume of water used, or kg/l. Thus recirculating aquaculture systems (RAS) are designed to minimise or reduced dependence on water usage.

Recirculating aquaculture systems are not very common in many areas. Even though they may not be economical for some species and location, they do provide a method of choice under certain conditions (Piedrahita 2003). These conditions are the real and perceived environmental benefits and the optimum utilisation of land and water. A recirculating aquaculture facility reduces water demands and discharges by reconditioning of water (Goldburg *et al.*, 2001). Better food conversions are achievable with a recirculating aquaculture system (RAS) which means less waste is generated by the feed (Lorsordo *et al.*, 1998). RAS systems have the advantage of temperature control which opens the door for year around production (Funge-Smith and Phillips, 2001).

General Functions of a Recirculating System

Generally, an efficient recirculating system should aerate the water, remove of particulate matter, remove waste ammonia and nitrite through biological filter and buffer pH. Removal of excess nutrients through biological filter or biofilter is the most important feature in any system. A biological filter facilitates nitrification, oxidises ammonia to nitrite and subsequently to less toxic nitrate, through nitrosomonas and nitrobactor bacteria. Solid filter performs the removal of suspended and settled particles through filter, settling chambers, etc. A pump is usually placed after biofilter to return water through system.

Biological filtration is the most common method of preventing the build up of toxic levels of ammonia in recirculating systems, while solid removal is comparatively easier and less expensive than other water treatment operations that may impact nutrients (Chen *et al.*, 1993). Basically, a biofilter is a substrate on which microorganisms grow. Media for biofilters can be virtually any substrate that provides maximum surface area for bacterial growth, for example shells, gravel, nettings, foam pads, etc.

Biofilters in Conventional Recirculating Aquaculture Systems

A conventional recirculating system generally uses four basic biofilter designs. They are trickling, submerged, updraft and rotating biodisc filters in addition to fluidised beds which can act as biofilters (Stickney 2000). The major function of a biofilter is the nitrification of ammonia. Recirculating biofilter technologies can be divided in two main categories: fixed film (attached growth) in which a media is provided for the microorganisms to attach and grow, and suspended growth in which the microorganisms are maintained in suspension. Most of the biofiltration on recirculating systems has been focused on aerobic, fixed film filters (Singh *et al.*, 1999; Malone and Beecher, 2000; Lekang and Kleppe, 2000; Sandu *et al.*, 2002) in which a substrate is provided for the growth of a biofilm that utilises oxygen to convert ammonia and nitrites to nitrates, and oxidise organic matter.

Recently, interest in other biofiltration alternatives, such as the suspended growth systems (Avnimelech, 1988; Van Wyk *et al.*, 1999) has increased. Fluidised-sand beds are an efficient, relatively compact, and cost-competitive technology for removing dissolved wastes from recirculating aquaculture systems, especially in relatively cool or coldwater applications that require maintaining consistently low levels of ammonia and nitrite. (Summerfelt, 2006).

The microorganism especially bacteria metabolise dissolved wastes that include dissolved organic compounds, ammonia, nitrite and nitrate. Many of these microorganisms live in biofilms that are located on surfaces within the biofilter and other substrates within the recirculating system (Sharrer *et al.*, 2004). These microorganisms are the integral part of the waste water treatment to manage water quality.

Need for Innovations in Biofilter Technology

Nitrogenous wastes in culture systems can be eliminated through nitrifying and denitrifying biofilters. While nitrifying filters are incorporated in most recirculating systems according to well-established, protocols, denitrifying filters are still under development. In addition to nitrate removal, denitrifying organisms are associated with other processes relevant to water quality control in aquaculture systems.

Denitrifiers also are associated with sulphur and phosphorus cycles in recirculating systems. Orthophosphate uptake by some denitrifiers takes place in excess of their metabolic requirements and may result in a considerable reduction of orthophosphate from the culture water. Information on nitrate removal in recirculating systems is limited to studies with small-scale experimental systems. Although studies on the application of denitrification in freshwater and marine recirculating systems were initiated some thirty years ago, a unifying concept for the design and operation of denitrifying biofilters in recirculating systems is lacking (Van Rijn *et al.*, 2006).

Recirculating aquaculture systems (RAS) are unable to compete directly with pond or flow-through systems producing freshwater commodity food fish. This is the main obstacle for a wider adoption of RAS technologies (Lorsordo and Westeman, 1994; Malone, 2002). Profitability of recirculating systems depends on the ability to manage the nutrients. Many innovations have occurred as result of research and development and increasing number of economically viable will be put into operation (Stickney 2000). Demand for cost effective biofilters will grow with expanded use of RAS. RAS systems will continue to grow in use both as a complement and as replacement of traditional ponds as the running cost come due to innovations and necessity to meet the bio-safety standards especially in grow-out facilities. Thus the innovations like the application of periphyton mats as biofilters in recirculating systems may provide an opportunity which is simple to operate and cost effective for the aquaculture

Periphyton-based Recirculatory Aquaculture Systems

Generally, periphyton refers to the assemblage of attached aquatic plants and animals on submerged substrates, including associated no-attached fauna (van Dam *et al.*, 2002). Majority of the active cells in a multi-species biofilm of a trickling filter were saprophytes and that 83-93 per cent of the biofilm biomass consisted of extra cellular polymeric substances (Hekmat *et al.*, 2004). The structure of a peiphyton mat develops into three dimensions in response to competition for space, nutrients and light. Algal periphyton assimilates ammonia and traps organic matter from water column and thus, as algal density increases, ammonia concentration decreases (Hargreaves, 1998).

Algae colonise the periphyton mat in higher densities than in water column. The mat which is formed by the attached organisms has lot of similarities with the biofilm used in conventional recirculating aquaculture systems (Verdegem *et al.*, 2005). Therefore periphyton mats could also carry out functions similar to that of biofilms. The concept of a periphyton based recirculating system is based on the function of algal turf scrubber (ATS) developed for maintaining water quality in aquatic microcosm (Adey, 1983). Water is passed over a grid which receives light so that algal mats are developed on these grids. These algal mats derive nutrients from the water and the excess algal mats are regularly scrubbed off from the grid leaving space for re-colonisation by algae. Lakatos *et al.*, (1999) confirmed role of periphyton-reed complex as biofilter in uptake of heavy metal at places impacted by high anthropogenic loadings.

Periphyton Complex as Biological Tool for Nutrient Removal

In RAS, the first goal is the removal of ammonia and

nitrite. Ammonia exists in two forms in water -unionised (NH_3) and ionised (NH_4^+). NH_3 is much more toxic than NH_4^+. The amount of ammonia present as NH_3 is dependent on temperature and pH of the water, increasing with both parameters. Fish release their nitrogenous wastes primarily as ammonia (NH_3) excreted across the gill membrane. Excess feed and other waste materials also add to the source of ammonia. To control ammonia levels in recirculating systems, extensive surface area is to be provided for the microorganism (bacteria) that oxidises ammonia into relatively harmless nitrate (NO_3-). In a periphyton mat, autotrophs, nitrifying bacteria and heterotrophs compete for space, nutrients and contribute to ammonia uptake (Verdegem *et al.*, 2005).

Inorganic nutrients are taken up from the overlying water to form new algal biomass (Sekar *et al.*, 2002). The potential of periphyton mat to remove ammonia was demonstrated by Azim *et al.*, (2002) in periphyton based culture ponds where bamboos were introduced as substrates at a density of 1 m^2 of submerged substrate for each m^2 of pond surface. The ammonia concentration of 0.56 mg/l in substrate ponds was significantly lower than 0.95 mg/l in substrate free pond.

The results of experiments on the efficiency of periphyton communities for nutrient removal from polluted streams in a continuous flow-through confirmed that periphyton communities are a useful means of nutrient removal from polluted streams (Vymazal, 1988). The artificial stream (5 m x 0.7 m x 0.5 m) was made of wood, with silon (a kind of nylon) screens, as a substratum for periphyton growth. During two field experiments a marked elimination of nitrogen and phosphorus was demonstrated. The maximum efficiency of ammonium and orthophosphate removal was 80 per cent and 70 per cent, respectively. Organic removal reached 35 per cent (C.O.D) and 54 per cent (B.O.D). Phosphorus uptake by the calcareous periphyton community in the oligotrophic Florida Everglades was reported by Scinto *et al.*, (2003).

The treatment performance of a 6 m wide by 67 m long

by 0.8 m deep, baffled sedimentation basin receiving rainbow trout *(Oncorhynchus mykiss)* raceway effluent was evaluated with and without the installation of artificial substrates which are commercially marketed as "Aquamats" (Nathan *et al.*, 2006). Treatment efficiency was determined using normal rearing condition effluent loading versus cleaning and harvesting events. Total suspended solids (TSS) removal for the total basin averaged 79 per cent and 71 per cent during normal rearing conditions, as compared to 92 per cent and 79 per cent during cleaning and harvesting operations, when the Aquamats were installed versus removed, respectively.

In most of the RAS, the waste water from the culture chambers passes through settling chambers before entering the biofilter and another settling chamber after leaving biofilter. A settling chamber is simply a tank in which the movement of water is sufficiently slow that solids will settle. Other methods for solid removal include mechanical Alteration through sand or a hydrocyclone device. Solid removal is important because it cuts down on the loading of the biofilter. Suspended solids are trapped in the periphyton mat, which also took up ammonia and nitrate, produced oxygen, broke down organic matter and increased nitrification. Nitrifying bacteria can also colonise the surface of the substrate that are located in well oxygenated water column rather than at anoxic water sediment. Therefore, periphytic biofilms on the substrate act as biofilter (Ramesh *et al.*, 1999). Compared to biofilms, periphyton mats may have the advantage of in situ oxygen production by the algae that are present (Verdegem *et al.*, 2005).

In the conventional RAS, considerable amount of nitrogen and phosphorous are still discharged thereby polluting the environment (Verdegem *et al.*, 2005). However these nutrients may be converted into easily harvestable plant materials in the form of algal mats in periphyton based RAS which can be harvested and utilised for various purposes. Algal turf scrubbing is a novel technology for the treatment of agricultural runoff and eutrophic lake water and may have application for wastewater treatment. The algal turf scrubber (ATS) consists of a natural, mixed assemblage of attached periphyton, microalgae and bacteria which colonise an inclined floway over which wastewater flows in a series of pulses. A large-scale ATS (152.4 m long 6.5 m wide) was constructed in Patterson, California and tested in conjunction with UV disinfection over one year for its ability to treat secondary effluent from an evaporation pond. The hydraulic loading rate of the wastewater was varied between 436 and 1226 m^3 per day and various operational parameters were tested.

The biomass was mechanically harvested from the floway at one or two week intervals depending upon the season. Based on the mean percentage of phosphorous (2.1 %) in the harvested solids and the mean productivity (35 g m^2 d^{-1}), the yearly mean removal of phosphorus was 0.73 ± 0.28 g m^{-2} d^{-1}. An inverse relationship was found between reduction in hydraulic loading rate and increase in pH, phosphorus removal and hardness reduction by the ATS. This indicated that pH mediated precipitation probably accounts for much of the phosphorus removal by the ATS and for the high mean phosphorus content of the harvested solids. Measurement of nutrient concentrations in influent and effluent of the ATS over 24 hours showed that at night phosphorus removal declined. These results indicate the potential of the ATS for phosphorus removal from wastewaters and suggest that removal may be easily controlled by altering the hydraulic loading rate (Craggs *et al.*, 1996).

Sereti *et al.*, (2004) compared the traditional system with trickling biofilter with a system equipped with an ATS for water treatment. The periphyton reactor was 1.6 m2 stainless steel grid which would support lab scale systems of 70 1 tanks. Water from fish tank through tipping bucket provided a constant wave action across the grid. The growing area was illuminated by metal halide lamps of 400W and pH was maintained between 7.0 and 7.5 with temperature at 27-28°C. Fish tanks were stocked with Nile tilapia at 25 kg/m^3 biomass.

The harvested fish constituted a biomass of 74 kg/m^3. The specific growth rate of 1.03 per cent body weight per day in periphyton-based RAS was better than 0.98 per cent in conventional RAS. Better FCR of 0.94 as compared 1.04 in conventional RAS showed the presence of additional food materials in form of algal mats which gave less than 1.0 FCR. Nitrogen and Phosphorus retention were 42 per cent and 58 per cent in periphyton based system while these were 30 per cent and 47 per cent in conventional RAS. The above model clearly indicated the feasibility of using periphyton mat as biofilter as well as the cost effectiveness due to better food conversion ratio.

Conclusion

Recirculating systems are designed to achieve overall environmental control in culture systems. Water quality control in RAS is attained through water exchange and water treatment. Therefore, the design of RAS should not only ensure optimum culture environment but also to reduce the operational cost that are still very high which hamper the use of these systems (Waller 2001). Good water quality can be managed in recirculating systems through effective operation of the system components and by avoiding water exchanges.

Given increasing land prices, water shortages that are occurring and projected to occur in the future, and the increasing governmental regulation on the effluents from aquaculture facilities, recirculating systems which will be cost effective will be needed if aquaculture is to grow in face of fierce competition for land water resources. In these contexts, the development of periphyton based recirculating systems seems promising in future.

5

Waste Management

Waste management is the collection, transport, processing, recycling or disposal of waste materials, usually ones produced by human activity, in an effort to reduce their effect on human health or local aesthetics or amenity. A subfocus in recent decades has been to reduce waste materials' effect on the natural world and the environment and to recover resources from them.

Waste management can involve solid, liquid or gaseous substances with different methods and fields of expertise for each.

Waste management practices differ for developed and developing nations, for urban and rural areas, and for residential, industrial, and commercial producers. Waste management for non-hazardous residential and institutional waste in metropolitan areas is usually the responsibility of local government authorities, while management for non-hazardous commercial and industrial waste is usually the responsibility of the generator.

Waste Management Concepts

There are a number of concepts about waste management, which vary in their usage between countries or regions.

Waste Hierarchy

The waste hierarchy refers to the "3 Rs" reduce, reuse and recycle, which classify waste management strategies according to their desirability.

The waste hierarchy has taken many forms over the past decade, but the basic concept has remained the cornerstone of most waste minimisation strategies. The aim of the waste hierarchy is to extract the maximum practical benefits from products and to generate the minimum amount of waste.

Some waste management experts have recently incorporated a 'fourth R': "Re-think", with the implied meaning that the present system may have fundamental flaws, and that a thoroughly effective system of waste management may need an entirely new way of looking at waste. Some "re-think" solutions may be counter-intuitive, such as cutting fabric patterns with slightly more "waste material" left — the now larger scraps are then used for cutting small parts of the pattern, resulting in a decrease in net waste. This type of solution is by no means limited to the clothing industry. *Source reduction* involves efforts to reduce hazardous waste and other materials by modifying industrial production. Source reduction methods involve changes in manufacturing technology, raw material inputs, and product formulation. At times, the term "pollution prevention" may refer to source reduction.

Another method of source reduction is to increase incentives for recycling. Many communities in the United States are implementing variable rate pricing for waste disposal (also known as Pay As You Throw - PAYT) which has been effective in reducing the size of the municipal waste stream.

Source reduction is typically measured by efficiencies and cutbacks in waste. *Toxics use reduction* is a more controversial approach to source reduction that targets and measures reductions in the upstream use of toxic materials. Toxics use reduction emphasises the more preventive aspects of source reduction but, due to its emphasis on toxic chemical inputs, has been opposed more vigorously by chemical manufacturers. *Toxics use reduction* programmes have been set up by legislation in some states, e.g., Massachusetts, New Jersey and Oregon.

Extended Producer Responsibility

Extended Producer Responsibility (EPR) is a strategy designed to promote the integration of environmental costs associated with products throughout their life cycles into the market price of the products (OECD 1999).

Extended producer responsibility imposes accountability over the entire life cycle of products and packaging introduced on the market. This means that firms, which manufacture, import and/or sell products and packaging, are required to be financially or physically responsible for such products after their useful life. They must either take back spent products and manage them through reuse, recycling or in energy production, or delegate this responsibility to a third party, a so-called *producer responsibility organisation* (PRO), which is paid by the producer for spent-product management. In this way, EPR shifts responsibility for waste from government to private industry, obliging producers, importers and/or sellers to internalise waste management costs in their product prices (Hanisch, 2000).

EPR has been implemented in many forms, which may be classified into three major approaches:

- Regulatory
- Negotiated
- Voluntary

It is perhaps because of the tendency of economic policy in market-driven economies not to interfere with consumers' preferences that the producer-centric representation is the dominant form of viewing the environmental impacts of industrial production: In statistics on energy, emissons, water, etc., impacts are almost always presented as attributes of industries ('on-site' or 'direct' allocation) rather than as attributes of the supply chains of products for consumers. On a smaller scale, most existing schemes for corporate sustainability reporting include only impacts that arise out of operations controlled by the reporting company, and not supply-chain impacts (World Business Council on Sustainable Development and World Resources Institute 2001). According to this world view, "upstream and downstream [environmental] impacts are allocated to their immediate producers. The institutional setting and the different actors' spheres of influence are not reflected."

On the other hand, a number of studies have highlighted that final consumption and affluence, especially in the industrialised world, are the main drivers for the level and growth of environmental pressure. Even though these studies provide a clear incentive for complementing producer-focused environmental policy with some consideration for consumption-related aspects, demand-side measures to environmental problems are rarely exploited.

The nexus created by the different views on impacts caused by industrial production is exemplified by several contributions to the discussion about producer or consumer responsibility for greenhouse gas emissions. Emissions data are reported to the IPCC as contributions of producing industries located in a particular country rather than as embodiments in products consumed by a particular population, irrespective of productive origin. However, especially for open economies, taking into account the greenhouse gases embodied in internationally traded commodities can have a considerable influence on national greenhouse gas balance sheets. Assuming consumer responsibility, exports have to be subtracted from, and imports added to national greenhouse gas inventories. In Denmark for example, Munksgaard and Pedersen (2001) report that a significant amount of power and other energy-intensive commodities are traded across Danish borders, and that between 1966 and 1994 the Danish foreign trade balance in terms of CO_2 developed from a 7 Mt deficit to a 7 Mt surplus, compared to total emissions of approximately 60 Mt. In particular, electricity traded between Norway, Sweden and Denmark is subject to large annual fluctuations due to varying rainfall in Norway and Sweden. In wet years Denmark imports hydroelectricity whereas electricity from coal-fired power plants is exported in dry years. The official Danish emissions inventory includes a correction for electricity trade and thus applies the consumer responsibility principle (Danish Environmental Protection Agency 1998).

Similarly, at the company level, "when adopting the concept of eco-efficiency and the scope of an environmental management system stated in for example ISO 14001, it is insufficient to merely report on the carbon dioxide emissions limited to the judicial borders of the company." "Companies must recognise their wider responsibility and manage the entire life-cycle of their products. Insisting on high environmental standards from suppliers and ensuring that raw materials are extracted or produced in an environmentally conscious way provides a start." A life-cycle perspective is also taken in Extended Producer Responsibility (EPR) frameworks: "Producers of products should bear a significant degree of responsibility (physical and/or financial) not only for the environmental impacts of their products downstream from the treatment and disposal of their product, but also for their upstream activities inherent in the selection of materials and in the design of products." "The major impetus for EPR came from northern European countries in the late 1980s and early 1990s, as they were facing severe landfill shortages. [... As a result,] EPR is generally applied to post-consumer wastes which place increasing physical and financial demands on municipal waste management."

EPR has rarely been consistently quantified. Moreover, applying conventional life cycle assessment, and assigning environmental impacts to producers and consumers can lead to double-counting. Using input-output analysis, researchers have attempted for decades to account for both producers and consumers in an economy in a consistent way. Gallego and Lenzen (2005) demonstrate and discuss a method of consistently delineating producers' supply chains, into mutually exclusive and collectively exhaustive responsibilities to be shared by all agents in an economy. Their method is an approach to allocating responsibility across agents in a fully interconnected circular system. Upstream and downstream environmental impacts are shared between all agents of a supply chain-producers and consumers.

Product Stewardship

Product stewardship is a concept whereby environmental protection centres around the product itself, and everyone involved in the lifespan of the product is called upon to take up responsibility to reduce its environmental impact. For manufacturers, this includes planning for, and if necessary, paying for the recycling or disposal of the product at the end of its useful life. For retailers and consumers, this means taking an active role in ensuring the proper disposal or recycling of an end-of-life product.

Those who advocate it are concerned with the later phases of product life cycle and the comprehensive outcome of the whole production process. It is considered a prerequisite to a strict service economy interpretation of (fictional, national, legal) "commodity" and "product" relationships.

The most familiar example is the container deposit charged for a deposit bottle. One pays a fee to buy the bottle, separately from the fee to buy what it contains. If one returns the bottle, the fee is returned, and the supplier must return the bottle for reuse or recycling. If not, one has paid the fee, and presumably this can pay for landfill or litter control measures that dispose of say a broken bottle. Also, since the same fee can be collected by anyone finding and returning the bottle, it is common for people to collect these and return them as a means of surviving. This is quite common for instance among homeless people in US cities.

However, the principle is applied very broadly beyond bottles to paint and automobile parts such as tires. When purchasing paint or tires in many places, one simultaneously pays for the disposal of the toxic waste they become. In some countries, such as Germany, law requires attention to the comprehensive outcome of the whole extraction, production, distribution, use and waste of a product, and holds those profiting from these legally responsible for any outcome along the way. This is also the trend in the UK and EU generally. In the United States, predictably, the issue has been confronted via class action lawsuits that attempt to hold companies liable for the environmental impact of their products. Thus far, such as litigation or proposed accounting reforms such as full cost accounting have not gained much traction for the product stewardship concept in the United States beyond the realm of academe and corporate public relations (derisively referred to as greenwashing).

The demand-side approach ethical consumerism, supported by consumer education and information about environmental impacts, may approach some of the same outcomes as product stewardship.

Extended Producer Responsibility

Product Stewardship is often used interchangeably with Extended Producer Responsibility, a similar concept. However, there are distinct differences between the two, as suggested by the semantics of the different terms used.

While both concepts bring the onus of waste management for end-of-life products from the government to the manufacturers, Product Stewardship further extends this responsibility to everyone involved in the life-cycle of the product. This includes not only the manufacturers, but also the retailers, consumers and recyclers as well.

Polluter Pays Principle

The Polluter Pays Principle is a principle in international environmental law where the polluting party pays for the damage done to the natural environment. It is regarded as a regional custom because of the strong support it has received in most Organisation for Economic Cooperation and Development (OECD) and European Community (EC) countries. International environmental law itself mentions little about the principle.

Polluter Pays is also known as Extended Polluter Responsibility (EPR). This is a concept that was probably first described by the Swedish Government in 1975. EPR seeks to shift the responsibility dealing with waste from governments to the entities producing it. In effect, it internalises the cost of waste disposal into the cost of the product, theoretically meaning that the producers will improve the waste profile of their products, thus decreasing waste and increasing possibilities for reuse and recycling.

OECD defines EPR as: "a policy in which the producer's financial and/or physical responsibility for a product is extended to the post-consumer stage of the product's life cycle. It specifically focuses on reducing the environmental impacts of a product at the post-consumer phase. There are two key features to an EPR policy: the responsibility for a product at its post— consumption phase is shifted upstream in the production-consumption chain, to the producer; and— it provides incentives to producers to incorporate environmental considerations into the design of their products."

Ecotaxes (as the Gas Guzzler Tax) are promoted as a way to implement the polluter pays principle. There are also "polluter pays" fines, i.e. in US Corporate Average Fuel Economy (CAFE).

Collection Methods

Collection methods vary widely between different countries and regions, and it would be impossible to describe them all. For example, in Australia most urban domestic households have a 240 litre (63.4 gallon) bin that is emptied weekly by the local council. In Europe and a few other places around the world, a few communities use a proprietary collection system known as

Envac, which conveys refuse via underground conduits using a vacuum system. Roosevelt Island has had this system since 1975. In Canadian urban centres curbside collection is the most common method of disposal, whereby the city collects waste and/or recyclables and/or organics on a scheduled basis. In rural areas people usually dispose of their waste at transfer stations. Waste collected is then transported to a regional landfill.

Many areas, especially those in less developed countries, do not have a formal waste-collection system in place.

Disposal methods also vary widely. In Australia, the most common method of disposal of solid waste is in landfill sites, as it is a large country with a low-density population. By contrast, in Japan it is more common for waste to be incinerated, because the country is smaller and land is scarce.

Landfill

A landfill, also known as a *dump* (and historically as a *midden*), is a site for the disposal of waste materials by burial and is the oldest form of waste treatment. Historically, landfills have been the most common methods of organised waste disposal and remain so in many places around the world.

Landfills may include internal waste disposal sites (where a producer of waste carries out their own waste disposal at the place of production) as well as sites used by many producers. Many landfills are also used for other waste management purposes, such as the temporary storage, consolidation and transfer, or processing of waste material (sorting, treatment, or recycling).

A landfill also may refer to ground that has been filled in with soil and rocks instead of waste materials, so that it can be used for a specific purpose, such as for building houses. Unless they are stabilised, these areas may experience severe shaking or liquefaction of the ground in a large earthquake.

Types and Construction Methods

Most modern landfills are classified according to the type(s) of waste material disposed of into them. Landfills can be engineered to a high standard in order to contain liquid leachate or landfill gas produced by decomposing organic waste. Modern landfills generally require a minimum of one landfill liner, consisting of a layer of compacted clay with a minimum required thickness and a maximum allowable hydraulic conductivity. Others also require the addition of one or more layers of impermeable membrane, such as high-density polyethylene (HDPE) together with geotextile. Various final cover systems are used to 'cap' landfills (such as clay or topsoil), depending on the type of wastes present within the landfill. Landfills, based on the waste type that is disposed within them, may be classified as:

1. *Hazardous Waste Landfill:* waste disposal units constructed to specific design criteria and which receive wastes meeting the local definition of hazardous waste. These landfills are generally constructed to be secure repositories for material that presents a serious hazard to human health, such as high-level radioactive waste. They are restricted, by permit or law, to the types of waste that they may handle (chemical vs. radioactive, liquid vs. dry). Double liner systems are the norm for hazardous waste landfills. Deep geological repository of high level radioactive waste is not generally classified as landfilling.
2. *Sanitary Landfills:* also called modern, engineered or secure landfills, these usually have physical barriers such as liners and leachate collection systems, and procedures to protect the public from exposure to the disposed wastes. The term sanitary landfill normally refers to those where municipal solid waste is disposed of, as well as other wastes high in organic material. In some countries, all landfills are sanitary landfills.
3. *Inert Waste Landfill:* waste disposal units that receive wastes which are chemically and physically stable and do not undergo decomposition, such as sand, bricks, concrete or gravel.

4. *Dumps:* also simply called landfills, dumps are landfills that are not engineered with the special protective measures required by sanitary landfills. They are most common in rural, remote, and developing areas. Many jurisdictions prohibit the use of non-sanitary landfills for the disposal of municipal solid waste. Other jurisdictions that do allow dumps may require them to be constructed according to some engineering standard to mitigate the risk for environmental contamination, such as by limiting the slope, requiring compaction, or ensuring that the cell is high enough above the groundwater table.

Subsystems

A typical landfill consists of subsystems such as the:

- Landfill liner
- Leachate collection and management system
- Landfill gas management system
- Landfill gas monitoring and leachate monitoring systems
- Road network
- Drainage system
- Final landfill cap

Their function is to secure the normal landfill operations and to control the anticipated emissions generated mainly by the decomposition of organic matter, such as leachate and landfill gas.

Site Construction Requirements

The construction of a landfill requires a staged approach. Landfill designers are primarily concerned with the viability of a site. To be commercially and environmentally viable a landfill must be constructed in accord with specific requirements, which are related to:

- Location
  - Easy access to transport by road

- Transfer stations if rail network is preferred
- Land value
- Cost of meeting government requirements, such as the Environment Agency in England and Wales
- Location of community served
- Type of construction (more than one may be used at single site)
  ⇒ *Pit*: Filling existing holes in the ground, typically left behind by mining
  ⇒ *Canyon:* Filling in naturally occurring valleys or canyons
  ⇒ *Mound:* Piling the waste up above the ground

- Stability
  - Underlying geology
  - Nearby earthquake faults
  - Water table
  - Location of nearby rivers, streams, and flood plains
- *Capacity:* The available voidspace must be calculated by comparison of the landform with a proposed restoration profile.

This calulation of capacity is based on,

- Density of the wastes
- Amount of intermediate and daily cover
- Amount of settlement that the waste will undergo following tipping
- Thickness of capping
- Construction of lining and drainage layers

- Protection of soil and water through:
  - Installation of liner and collection systems
  - Storm water control
  - Leachate management
  - Landfill gas management

- Nuisances and hazards management
- Costs
  - Feasiability studies
  - Site after care
  - Site investigations (costs involved make small sites uneconomic)

Operations

Typically, in non- hazardous waste landfills, in order to meet pre-defined specifications, techniques are applied by which the wastes are:

1. Confined to as small an area as possible.
2. Compacted to reduce their volume.
3. Covered (usually daily) with layers of soil.

During landfill operations the waste collection vehicles are weighed at a weigh-bridge on arrival and their load is inspected for wastes that do not accord with the landfill's waste acceptance criteria. Afterwards, the waste collection vehicles use the existing road network on their way to the tipping face or working front where they unload their load. After loads are deposited, compactors or dozers are used to spread and compact the waste on the working face. Before leaving the landfill boundaries, the waste collection vehicles pass through the wheel cleaning facility. If necessary, they return to weighbridge in order to be weighed without their load. In addition to trucks, some landfills may be equipped to handle railroad containers. The use of 'rail-haul' permits landfills to be located at more remote sites, without the problems associated with many truck trips.

Typically, in the working face, the compacted waste is covered with soil daily. Alternative waste cover materials are several sprayed on foam products and temporary blankets. Blankets can be lifted into place with tracked excavators and then removed the following day prior to waste placement. Chipped wood and chemically 'fixed' biosolids, may also be used as an alternate daily cover. The space that is occupied daily by the compacted waste and the cover material is called daily cell. Waste compaction is critical to extending the landfill life. Factors such as waste compressibility, waste layer thickness and the number of passes of the compactor over the waste affect the waste densities.

Land Reclamation

As human overcrowding of developed areas intensified during the 20th century, it has become important to develop land reuse strategies for completed landfills. Some of the most common usages are for parks, golf courses and other sports fields. Increasingly, however, office buildings and industrial uses are made of a completed landfill. In these latter uses, methane capture is customarily carried out to minimise explosive hazard within the building.

An example of a Class A office building constructed over a landfill is the Dakin Building at Sierra Point, Brisbane, California. The underlying fill was deposited from 1965 to 1985, mostly consisting of construction debris from San Francisco and some municipal wastes. Aerial photographs prior to 1965 show this area to be tidelands of the San Francisco Bay. A clay cap was constructed over the debris prior to building approval.

Another strategy for landfill reclamation is the incineration of landfill trash at high temperature via the plasma-arc gasification process, which is currently used at two facilities in Japan, and will be used at a planned facility in St. Lucie County, Florida.

Impacts

A number of adverse impacts occur from landfill operations. These impacts can vary: fatal accidents (e.g., scavengers buried under waste piles), infrastructure damage (e.g., damage to access roads by heavy vehicles), pollution of the local environment (such as contamination of groundwater and/or aquifers by leakage and residual soil contamination after landfill closure), offgassing of methane generated by decaying organic wastes; harbouring of disease vectors such as rats and flies, particularly from improperly operated landfills, which are common in Third-world countries injuries to wildlife and simple nuisance problems (e.g., dust, odour, vermin, or noise pollution).

Environmental noise and dust are generated from vehicles accessing a landfill as well as from working face operations. These impacts are best to intercept at the planning stage where access routes and landfill geometrics can be used to mitigate such issues. Vector control is also important, but can be managed reasonably well with the daily cover protocols.

Most modern landfills are operated with controls to manage problems such as these. Analysis of common landfill operational problems are available in.

Some local authorities have found it difficult to locate new landfills. These authorities may charge a fee or levy in order to discourage waste and/or recover the costs of site operations. Some landfills are operated for profit as commercial businesses. Many landfills, however, are publicly operated and funded.

Regional Practice

United Kingdom: Landfilling practices in the UK have had to change in recent years to meet the challenges of the European Landfill Directive. The UK now imposes landfill tax upon biodegradable waste which is landfilled. In addition to this the Landfill Allowance Trading Scheme has been established for local authorities to trade landfill quotas.

United States: In the US, landfills are regulated by the state's environmental agency that establishes minimum guidelines; however, none of these standards may fall below those set by the US EPA.

The Fresno Municipal Sanitary Landfill, opened in Fresno, California in 1937, is considered to have been the first modern, sanitary landfill in the United States, innovating the techniques of trenching, compacting, and the daily covering of waste with soil. It has been designated a National Historic Landmark, underlining the significance of waste disposal in urban society.

Before the advent of modern landfills in America, most Americans lived in sparsely populated rural farming communities and most burned their garbage. Due to environmental and safety concerns, burning garbage by civilians has been outlawed by most municipalities and can only be performed by landfill managers or people who have obtained permits from the municipality.

The Resource Conservation and Recovery Act is a US federal law that is designed to protect the public from harm caused by waste disposal.

US landfills consist of 40 per cent to 50 per cent paper waste, 20 per cent to 30 per cent construction debris, and 1.2 per cent disposable diapers.

Alternatives

The obvious alternative to landfills are waste reduction and recycling strategies. Secondary to not creating waste, there are various alternatives to landfills. In the late 20th century, alternative methods to waste disposal to landfill and incineration have begun to gain acceptance. Anaerobic digestion, composting, mechanical biological treatment, pyrolysis and gasification have all began to establish themselves in the market.

In recent years, some countries, such as Germany, Austria, and Switzerland, have banned the disposal of untreated waste in landfills. In these countries, only the ashes from incineration or the stabilised output of mechanical biological treatment plants may still be deposited.

Incineration

Incineration is a waste treatment technology that involves the combustion of waste at high temperatures. Incineration and other high temperature waste treatment systems are described as "thermal treatment". Incineration of waste materials converts the waste into heat (that can be used to generate electricity), gases and particulates of combustion, and a smaller amount of solid residue (ash). The gaseous and particulate emissions are dispersed in the atmosphere if the incinerator is of the open variety, but new designs store the emissions in underground reservoirs such as empty oil wells. Incinerators that burn municipal waste are often referred to as MSWIs: Municipal Solid Waste Incinerators.

A waste-to-energy plant (WtE) is a modern term for an incinerator that burns wastes in high-efficiency furnace/boilers to produce steam and/or electricity and incorporates modern air pollution control systems and continuous emissions monitors. This type of incinerator is sometimes called an energy-from-waste (EfW).

Incineration functions as an alternative to landfilling and biological treatment methods such as composting and anaerobic digestion.

Incineration has particularly strong benefits for the treatment of certain waste types in niche areas such as clinical wastes and certain hazardous wastes where pathogens and toxins must be destroyed by high temperatures.

Modern incinerators are very different from the incinerators that were commonly used until a few decades ago. Old-type incinerators usually did not include a materials separation to remove hazardous or recyclable materials before burning, and tended to risk the health of the plant workers and the nearby residents, and most of them did not generate electricity.

The potential of electricity generation using municipal waste combustion and other non-thermal methods of waste-based energy such as anaerobic digestion are being increasingly looked at as a potential energy diversification strategy.

Incineration is particularly popular in countries such as Japan where land is a scarce resource. Sweden has been a leader in using the energy generated from incineration over the past 20 years. Denmark also extensively uses waste-to-energy incineration in localised combined heat and power facilities supporting district heating schemes.

Modern furniture factory sawdust incinerators need much attention as these have to handle resin powder and many flammable substances. Controlled combustion, burn back prevention systems are very essential as dust when suspended resembles the fire catch phenomenon of any liquid petroleum gas.

Technology

Types of Incinerators

An incinerator is a furnace for burning refuse, modern incinerators include pollution mitigation equipment such as flue-gas cleaning. There are various types of incinerator plant design:

- Simple
- Fixed or moving grate combustion
- Rotary-kiln
- Multiple/stepped hearth
- Fluidised bed

Simple Incinerators

The older and simpler kind of incinerator was a brick-lined cell with a metal grate over a lower ash pit, with one opening in the top or side for loading and another opening in the side for removing incombustible solids called clinkers. Many small incinerators formerly found in apartment houses have now been replaced by waste compactors.

Fixed or Moving Grate Combustion

These are large fixed hearth incinerators, with a moving grate. The moving grate enables the movement of waste through the combustion chamber to be optimised to allow a more efficient burn. These incinerators are typically used for combustion of Municipal wastes.

Rotary-kiln

The rotary-kiln incinerator used by municipalities and by large industrial plants. This design of incinerators have

2 chambers a primary chamber and secondary chamber. The primary chamber in a rotary klin incinerator consist of an inclined refractory lined cylindrical tube. Movement of the cylinder on its axis facilitates movement of waste. In the primary chamber, there is conversion of solid fraction to gases, through volatilisation, destructive distillation and partial combustion reactions. The secondary chamber is necessary to complete gas phase combustion reactions

The clinkers spill out at the end of the cylinder. A tall flue gas stack, fan, or steam jet supplies the needed draft. Ash drops through the grate, but many particles are carried along with the hot gases. The particles and any combustible gases may be combusted in an "afterburner". To control air pollution, the combustion product gases are further treated with acid gas scrubbers to remove sulphuric acid and nitric acid emissions, and then routed through bag houses to remove particulates before the gases are released into the atmosphere.

Multiple-stepped Hearth

Waste is transported through the furnace by moving teeth mounted on a central rotating shaft.

Fluidised Bed

A strong airflow is forced through a sandbed The air seeps through the sand until a point is reached where the sand particles separate to let the air through and mixing and churning occurs, thus a fluidised bed is created and fuel and waste can now be introduced. The sand with the pre-treated waste and/ or fuel is kept suspended on pumped air currents and takes on a fluid-like character. The bed is thereby violently mixed and agitated keeping small inert particles and air in a fluid like state. This allows all of the mass of waste, fuel and sand to be fully ciculated through the furnace.

Use of Heat

The heat produced by the rotary-kiln incinerator can be used to generate steam which may then be used to drive an electrical generator. The typical range of net electrical energy that can be produced is about 500 to 600 kWh of per ton of waste incinerated. Thus, incinerating about 2,200 tons per day of waste will produce about 50 MW of electrical power.

Pollution

Incineration has a number of outputs such as the ash and the emission to the atmosphere of combustion product gases and particulate matter.

Modern incinerators emit less air pollution than coal plants, but more than natural gas plants. The US EPA has characterised modern incinerators as "producing electricity with less environmental impact than almost any other source of electricity". Germany's Ministry of the Environment in September 2005 published a report stating that modern incinerators reduce the amount of some atmospheric pollutants by substituting power produced by coal-fired plants with power from waste-fired plants. Link: Waste incineration a potential danger: bidding farewell to dioxin spouting.

Gaseous Emissions

The combustion product gases exhausted to the atmosphere by incineration are a source of concern. The main pollutants in the exhaust gases include acid gases such as hydrogen chloride, sulphur dioxide, nitrogen oxides (referred to as NO_x), and carbon dioxide.

The most publicised concerns from environmentalists about the incineration of municipal solid wastes (MSW) involve the fear that it produces significant amounts of dioxin and furan emissions to the atmosphere. Dioxins and furans are considered by many to be serious health hazards. Older generation incinerators that were not equipped with modern gas cleaning technologies were indeed significant sources of dioxin emissions. Today, however, due to advances in emission control designs and stringent new governmental regulations, modern waste-to-energy incinerators emit virtually no dioxins.

The quantity of pollutants in the emissions from large-scale incinerators is reduced by a process known as scrubbing as well as other processes.

According to the US EPA, modern incinerators are no longer significant sources of dioxins and furans. In 1987, before the governmental regulations required the use of emission controls, there was a total of 10,000 grams of dioxin emissions from US waste-to-energy incinerators. Today, the total emissions from the 87 plants are only 10 grams, a reduction of 99.9 per cent. Backyard barrel burning of household and garden wastes, still allowed in some rural areas, generates 580 grams of dioxins yearly. Studies conducted by the US EPA demonstrate that the emissions from just one family using a burn barrel produces more emissions than a modern incinerator disposing of 200 tons of waste per day.

According to the 2005 report from the Ministry of the Environment of Germany, where there are 66 incinerators, "...whereas in 1990 one third of all dioxin emissions in Germany came from waste incineration plants, for the year 2000 the figure was less than 1 per cent. Chimneys and tiled stoves in private households alone discharge approximately twenty times more dioxin into the environment than waste incineration plants."

Solid Outputs

Incineration produces fly ash and bottom ash just as is the case when coal is combusted. The total amount of ash produced by municipal solid waste incineration ranges from 15 per cent to 25 per cent by weight of the original quantity of waste, and the fly ash amounts to about 10 per cent to 20 per cent of the total ash. The fly ash, by far, constitutes more of a potential health hazard than does the bottom ash because the fly ash contains toxic metals such as lead, cadmium, copper and zinc as well as small amounts of dioxins and furans. The bottom ash may or may not contain significant levels of health hazardous materials. In the United States, and perhaps in other countries as well, the law requires that the ash be tested for toxicity before disposal in landfills. If the ash is found to be hazardous, it can only be disposed of in landfills which are carefully designed to prevent pollutants in the ash from leaching into underground aquifers, but in testing over the past decade, no ash from a US modern waste-to-energy plant has ever been determined to be a hazardous waste; In the UK, fly ash is classed as hazardous and must be disposed of in a hazardous waste designated landfill but Incinerator Bottom Ash (IBA) is under investigation by the Environment Agency.

At present although some historic samples tested by the incinerator operators' group would meet the being ecotoxic criteria at present the EA say "we have agreed" to regard incinerator bottom ash as "non-hazardous" until the testing programme is complete.

Pollution Issues

Odour pollution can be a problem with old-style incinerators, but it is claimed that odours and dust are extremely well controlled in a modern incinerator. They receive and store the waste in an enclosed area with a negative pressure with the airflow being routed through a boiler which prevents unpleasant odours from escaping into the atmosphere. However, not all plants are implemented this way, resulting in complaints.

An issue that affects community relationships is the increased road traffic of waste collection vehicles to transport municipal waste to the incinerator. Due to this reason, most incinerators are located in industrial areas.

Incineration

Debate over Incineration: Use of incinerators for waste management is controversial. The debate over incinerators typically involves business interests (representing both waste generators and incinerator firms), government regulators, and local citizens who must weigh the economic appeal of local industrial activity with their concerns over health and environmental risk.

People and organisations professionally involved in this issue include the US Environmental Protection Agency (US EPA) and a great many local and national air quality regulatory agencies worldwide.

Argument for Incineration

- The concerns over the health effects of dioxin and furan emissions have been significantly lessened by advances in emission control designs and very stringent new governmental regulations that have resulted in large reductions in the amount of dioxins and furans emissions.
- Modern incinerators generate electricity and heat that can be sold to the regional electric grid and can sell steam to district heating systems or industrial customers.
- The ash residue remaining after combustion has been shown to be a non-hazardous solid waste that can be safely landfilled or possibly reused.
- In densely populated areas, finding space for additional landfills is becoming very difficult.
- Countries considered "green" such as Sweden, Denmark, Switzerland and Germany, among others, rely heavily on incinerators, with some countries using incineration to dispose of all of their post-recycling municipal waste.
- Incineration of medical waste produces an end product ash that is sterile and non-hazardous.

Argument against Incineration

- Modern incineration facilities cause environmental harm of a similar magnitude to a modern engineered landfill.

- There are still concerns by many about the health effects of dioxin and furan emissions into the atmosphere
- Incinerators also emit heavy metals such as mercury, lead and cadmium, which can be toxic at very minute levels.
- The expense of building and operating an incinerator.
- Although waste incineration can be used to generate energy, a portion of that energy is consumed by the use of scrubbers and other methods to clean up the exhaust gases.
- The end product ash must still be safely disposed of. SAARC (South Asian Association for Regional Cooperation) countries agree that incineration as well as unproven technologies such as Plasma, should not be considered as an option for the treatment of their municipal solid wastes for low calorific value and environmental pollution potential.

Trends in Incinerator Use

The history of municipal solid waste (MSW) incineration is linked intimately to the history of landfills and other waste treatment technology. The merits of incineration are inevitably judged in relation to the alternatives available. Since the 1970s, recycling and other prevention measures have changed the context for such judgements. Since the 1990s alternative waste treatment technologies have been maturing and becoming viable.

Incineration is a key process in the treatment of hazardous wastes and clinical wastes. It is often imperative that medical waste be subjected to the high temperatures of incineration to destroy pathogens and toxic contamination it contains.

Incineration in United States

The first full-scale waste-to-energy incineration facility in the US was the Arnold O. Chantland Resource Recovery Plant, built in 1975 located in Ames, Iowa. This plant is still in operation and produces refuse-derived fuel that is sent to local power plants for fuel.

Whereas the construction of modern incinerators plants has increased in Europe, Japan and China during the past 15 years, in the US there have been no new plants constructed since the early 1990s. Several older generation incinerators have been closed; of the 186 MSW incinerators in 1990, only 112 remained by 2003, and of the 6200 medical waste incinerators in 1988, only 115 remained in 2003. The main reasons for lack of activity have been:

- *Law:* Supreme Court's 1994 ruling in Carbone v. Town of Clarkson, which held that "flow control", or laws that direct where waste should be processed or disposed, violate the Commerce Clause. The effect of this ruling was to make it too risky for private developers to construct new incinerators.
- *Economics:* With the increase in the number of large inexpensive regional landfills and, up until recently, the relatively low price of electricity, incinerators were not able to compete for the 'fuel', i.e., waste. In Europe, with the ban on landfilling untreated waste, scores of modern design incinerators have been built in the last decade, with more under construction. Recently, a number of municipal governments have begun the process of contracting for the construction and operation of Incinerators. A number of Canadian cities are likewise working towards installation of incinerators.
- *Tax Policies:* Tax credits for plants producing electricity from waste were rescinded in the 1990s. In Europe, some of the electricity generated from waste is deemed to be from a 'Renewable Energy Source (RES)'. A new law granting tax credits for such plants was implemented in the US in 2004.
- *NIMBY:* Local opposition together with environmental activists have thwarted projects in major US cities.

- *Landfill Bias:* In Europe and Japan, the public understands the hazards of landfills, but this awareness is less developed in the US, where furthermore there is a lot of open land.

Despite these problems, there has been renewed interest in incineration in the US Canada and the UK. Projects to add capacity to existing plants are underway, and municipalities are once again evaluating the option of building modern plants rather than continue landfilling municipal wastes.

Incineration in United Kingdom

The technology employed in the UK waste management industry has been greatly lagging behind that of Europe due to the wide availability of landfills. The Landfill Directive set down by the European Union led to the Government of the United Kingdom imposing waste legislation including the landfill tax and Landfill Allowance Trading Scheme. This legislation is designed to reduce the release of greenhouse gases produced by landfills through the use of alternative methods of waste treatment. It is the UK Government's position that incineration will play an increasingly large role in the treatment of municipal waste and supply of energy in the UK.

Incineration in Developing Countries

Mismanagement of health care waste puts health care workers, patients and the community at risk from pathogens and from pollution due to burning in open pits or badly maintained incineration equipment. Primary health facilities require health care waste management systems to minimise the risk of contamination of patients, health workers and the general public from infectious waste.

In this regard, a Waste Disposal Unit (WDU) with Small Scale Incinerator (SSI), when used according to Best Practices, can be a cheap and comparatively less hazardous way of disposing of health care waste.

Mobile incinerators are becoming more widely used in

developing countries where the threat of avian influenza is high. Small incinerators can be quickly deployed to remote areas where an outbreak has occurred to dispose of infected animals quickly and without the risk of cross contamination, mobile incinerators, self-sufficient units are ideal for remote areas where fuel and electricity are not available.

Materials Recovery Facility

A materials recovery facility or materials reclamation facility (MRF — pronounced "murf") is a specialised plant that receives, separates and prepares recyclable materials for marketing to end-user manufacturers. Generally, there are two types—clean and dirty MRF's.

Clean MRF

A *clean MRF* accepts recyclable materials that have been collected in comingled wastes from kerbside collection separated at source from municipal solid waste generated by either residential or commercial sources. There are a variety of clean MRF's. The most common currently are 2-stream MRF's, where source-separated recyclables are delivered in the form of a mixed food and beverage container stream (typically glass, ferrous metal, aluminium and other non-ferrous metals, PET [No.1] and HDPE [No.2] plastics) and a mixed paper stream.

The composition of the mixed paper stream can vary considerably, depending on the marketing arrangements that are available to the MRF operator. A typical mixed paper stream will consist of old newspapers with their inserts, old magazines and kraft (brown) paper bags. Some MRF's are able to handle a mixed paper stream consisting of a wider variety of paper types that includes old corrugated, junk mail, telephone books and even paperboard.

Dirty MRF

A *dirty MRF* accepts a mixed solid waste stream and then proceeds to separate out designated recyclable materials through a combination of manual and mechanical sorting. The sorted recyclable materials may undergo further processing required to meet technical specifications established by end-markets while the balance of the mixed waste stream is sent to a disposal facility such as a landfill.

The percentage of residuals (unrecoverable recyclable or non-programme materials) from a properly operated clean MRF supported by an effective public outreach and education programme should not exceed 10 per cent by weight of the total delivered stream and in many cases it can be significantly below 5 per cent. A dirty MRF recovers between 5 per cent and 45 per cent of the incoming material as recyclables, then the remainder is landfilled or otherwise disposed.

New mechanical biological treatment technologies such as the ArrowBio Process are now beginning to utilise *wet MRFs*. This combines a dirty MRF with water which acts to density separate and clean the output streams. It also hydrocrushes and dissolves biodegradable organics in solution to make them suitable for anaerobic digestion.

Recycling

Recycling is the reprocessing of materials into new products. Recycling prevents useful material resources being wasted, reduces the consumption of raw materials and reduces energy usage, and hence greenhouse gas emissions, compared to virgin production. Recycling is a key concept of modern waste management and is the third component of the waste hierarchy.

Recyclable materials, also called "recyclables" or "recyclates", may originate from a wide range of sources including the home and industry. They include glass, paper, aluminium, asphalt, iron, textiles and plastics. Biodegradable waste, such as food waste or garden waste, is also recyclable with the assistance of microorganisms through composting or anaerobic digestion.

Recyclates need to be sorted and separated into material types. Contamination of the recylates with other materials must be prevented to increase the recyclates value and facilitate easier reprocessing for the ultimate recycling facility. This sorting can be performed either by the producer of the waste or within semi- or fully-automated materials recovery facilities.

There are two common household methods of helping increase recycling. In curbside collection (UK: kerbside collection) consumers leave presorted materials for recycling in front of their property, typically in boxes or sacks to be collected by a recycling vehicle. Alternatively, with a "bring" or carry-in system, the householder may take the materials to collection points, such as transfer stations or civic amenity sites, where recyclates are placed into recycling bins based on the type of material.

Recycling does not include reuse where items retain their existing form for other purposes without the need for reproducing. That said it does include regiving where simply ownership changes as items are gifted from one owner to another.

Recycling has been a common practice throughout human history. In pre-industrial times, scrap made of bronze and other precious metals were collected in Europe and melted down for perpetual reuse, and in Britain dust and ash from wood and coal fires was downcycled as a base material in brick making. The main driver for these types of recycling was the economic advantage of obtaining recycled feedstock instead of acquiring virgin material, as well as a lack of public waste removal in ever more-populated sites.

Paper recycling began in Britain in 1921, when the British Waste Paper Association (now Confederation of Paper Industries) was established to encourage trade in waste paper recycling.

Resource shortages caused by the world wars, and other such world-changing occurrences greatly encouraged recycling. Massive government promotion campaigns were carried out in World War II in every country involved in the war, urging citizens to donate metals and conserve fibre, as a matter of significant patriotic importance. Resource conservation programmes established during the war were continued in some countries without an abundance of natural resources, such as Japan, after the war ended.

The next big investment in recycling occurred in the 1970s, due to rising energy costs (recycling aluminium uses only 5 per cent of the energy required by virgin production; glass, paper and metals have less dramatic but very significant energy savings when recycled feedstock is used). The passage of the Clean Water Act of 1977 in the USA created strong demand for bleached paper (office paper whose fibre has already been bleached white increased in value as water effluent became more expensive).

In 1973, the city of Berkeley, California began one of the first curbside collection programmes with monthly pick ups of newspapers from residences. Since then several countries have started and expanded various doorstep collection schemes.

In 1987, the Mobro 4000 barge hauled garbage from New York to North Carolina, where it had been denied. It was then sent to Belize, where it was denied as well. Finally, the barge returned to New York and the garbage was incinerated. The incident led to heated discussions about waste disposal and recycling.

One event that initiated recycling efforts occurred in 1989 when Berkeley banned the use of polystyrene packaging for keeping McDonald's hamburgers warm. One effect of this ban was to raise the ire of management at Dow Chemical, the world's largest manufacturer of polystyrene, which led to the first major effort to show that plastics can be recycled. By 1999, there were 1,677 companies in the USA alone involved in the post-consumer plastics recycling business.

Benefits

Recycling is beneficial in two ways: it reduces the inputs

(energy and raw materials) to a production system and reduces the amount of waste produced for disposal.

Some materials like aluminium can be recycled indefinitely as there is no change to the materials. Other recycled materials like paper require a percentage of raw materials (wood fibres) to be added to compensate for the degradation of existing fibres.

The resources being processed are purer, less energy is needed to process them and less energy is needed to transport from the place of extraction (e.g. bauxite/aluminium ore mines in Brazil or coniferous forests in Scandinavia).

This reduces the environmental, social, and usually the economic costs of manufacturing.

For example, bauxite mines in Brazil displace indigenous people, create noise pollution from blasting, machinery and transport, and create air pollution in the form of particulates (dust). The habitat loss and visual destruction is also negative both to the aesthetic qualities of the areas and the local environment. However, the mines do provide employment and revenue to the local population and economy, promoting development of the country as a whole.

Recycling aluminium saves 95 per cent of the energy cost of processing new aluminium because the melting temperature is reduced from 900°C to 600°C. It is by far the most efficient material to recycle.

The most commonly used methods for waste disposal (landfill, pyrolysis, incineration) may be environmentally damaging and unsustainable. Therefore any way to reduce the volume of waste being disposed in this fashion may be beneficial. The maximum environmental benefit is gained by waste minimisation (reducing the amount of waste produced), and reusing items in their current form such as refilling bottles.

Drawbacks

All recycling techniques consume energy for transportation

and processing, and some also use considerable amounts of water.

There may also be drawbacks with the collection methods associated with recycling. Increasing collections of separated wastes adds to vehicle movements and the production of carbon dioxide. This may be negated however by centralised facilities such as some advanced material recovery facilities of mechanical biological treatment systems for the separation of mixed wastes. However, this is almost never the case for urban areas, taking into account the massive number of workers, machines, and vehicles needed for the recycling process.

Perverse consequences from mercury recycling have been cited recently by the Wall Street Journal (April 20, 2006). The article traces mercury recovered from USA recycling programmes into sales of mercury for alluvial mining activities in Brazil. During the autumn of 2006, the EU banned the export of liquid mercury (Europe has no mercury mining, only recovery from recycling). A full life-cycle analysis prior to institution of recycling programmes may reduce the risk of unintended environmental consequences.

Recycling Techniques

Many different materials can be recycled but each type requires a different technique.

Aggregates and Concrete

Concrete aggregate collected from demolition sites is put through a crushing machine, often along with asphalt, bricks, dirt, and rocks. Smaller pieces of concrete are used as gravel for new construction projects. Crushed recycled concrete can also be used as the dry aggregate for brand new concrete if it is free of contaminants. This reduces the need for other rocks to be dug up, which in turn saves trees and habitats.

Batteries

The large variation in size and type of batteries makes their recycling extremely difficult: they must first be sorted into similar kinds and each kind requires an individual recycling process. Additionally, older batteries contain mercury and cadmium, harmful materials which must be handled with care.

Lead-acid batteries, like those used in automobiles, are relatively easy to recycle and many new lead-acid batteries contain a high percentage of recycled material.

Biodegradable Waste

Biodegradable waste can be recycled into useful material by biological decomposition. There are two mechanisms by which this can occur. The most common mechanism of recycling of household organic waste is home composting or municipal curbside collection of green wastes sent to large scale composting plants.

Alternatively organic waste can be converted into biogas and soil improver using anaerobic digestion. Here organic wastes are broken down by anaerobic microorganisms in biogas plants. The biogas can be converted into renewable electricity or burnt for environmentally friendly heating. Advanced technologies such as mechanical biological treatment are able to sort the recyclable elements of the waste out before biological treatment by either composting, anaerobic digestion or biodrying.

Electronics Disassembly and Reclamation

The direct disposal of electrical equipment — such as old computers and mobile phones — is banned in many areas due to the toxic contents of certain components. The recycling process works by mechanically separating the metals, plastics and circuit boards contained in the appliance. When this is done on a large scale at an electronic waste recycling plant, component recovery can be achieved in a cost-effective manner.

Electronic devices, including audio-visual components (televisions, VCRs, stereo equipment), mobile phones and other hand-held devices, and computer components, contain valuable elements and substances suitable for reclamation, including lead, copper, and gold. They also contain a plethora of toxic substances, such as dioxins, PCBs, cadmium, chromium, radioactive isotopes, and mercury. Additionally, the processing required to reclaim the precious substances (including incineration and acid treatments) release, generate and synthesise further toxic by-products.

In the United States, an estimated 70 per cent of heavy metals in landfills come from discarded electronics. Some regional governments are attempting to curtail the accumulation of electronics in landfills by passing laws obligating manufacturers and consumers to recycle these devices, but because in many cases safe dismantlement of these devices in accordance with first world safety standards is unprofitable, historically much of the electronic waste has been shipped to countries with lower or less rigorously-enforced safety protocols. Places like Guiyu, China dismantle tons of electronics every year, profiting from the sale of precious metals, but at the cost of the local environment and the health of its residents.

Mining to produce the same metals, to meet demand for finished products in the west, also occurs in the same countries, and the United Nations Conference on Trade and Development (UNCTAD) has recommended that restrictions against recycling exports be balanced against the environmental costs of recovering those materials from mining. Hard rock mining in the USA produces 45 per cent of all toxics produced by all USA industries (2001 US EPA Toxics Release Inventory).

Printer Ink Cartridges and Toners

Printer ink cartridges can be recycled. They are sorted into different brands and models which are then resold back to the companies that created these cartridges. The companies then refill the ink reservoir which can be sold back to consumers. Toner cartridges are recycled the same way as ink cartridges, using toner instead of ink. This method of recycling is highly efficient as there is no energy spent on melting and recreating the recycled object itself.

Ferrous Metals

Iron and steel are the world's most recycled materials, and among the easiest materials to recycle, as they can be separated magnetically from the waste stream. Recycling is via a steelworks: scrap is either remelted in an Electric Arc Furnace (90-100 per cent scrap), or used as part of the charge in a Basic Oxygen Furnace (around 25 per cent scrap). Any grade of steel can be recycled to top quality new metal, with no 'downgrading' from prime to lower quality materials as steel is recycled repeatedly. 42 per cent of crude steel produced is recycled material.

Non-ferrous Metals

Aluminium is shredded and ground into small pieces or crushed into bales. These pieces or bales are melted in an aluminium smelter to produce molten aluminium. By this stage the recycled aluminium is indistinguishable from virgin aluminium and further processing is identical for both.

Due to the high melting point of aluminium ore, large amounts of energy are required to extract aluminium from ore, making the environmental benefits of recycling aluminium enormous. Approximately 5 per cent of the CO_2 is produced during the recycling process compared to producing raw aluminium (and an even smaller percentage when considering the complete cycle of mining and transporting the aluminium). Also, as open-cut mining most often used for obtaining aluminium ore, mining destroys large sections of natural land. An aluminium can is 100 per cent recyclable every time it is recycled, it saves enough energy to watch television for about three hours (compared to mining and producing a new can).

Glass

Glass bottles and jars are accumulated via curbside collection schemes and bottle banks, where the glass may be sorted into colour categories. The collected glass *cullet* is taken to a glass recycling plant where it is monitored for purity and contaminants are removed. The cullet is crushed and added to a raw material mix in a melting furnace. It is then mechanically blown or moulded into new jars or bottles. Glass cullet is also used in the construction industry for aggregate and glassphalt. Glassphalt is a road-laying material which comprises around 30 per cent recycled glass. Glass can be recycled indefinitely as its structure does not deteriorate when reprocessed.

Paper

Recycled paper is made from waste paper, usually mixed with fresh wood pulp. If the paper contains ink, it must be deinked. This also removes fillers, clays, and fibre fragments.

Almost all paper can be recycled today, but some types are harder to recycle than others. Kraft paper, papers coated with plastic or aluminium foil, and papers that are waxed, pasted, or gummed are usually not recycled because the process is too expensive. Gift wrap paper also cannot be recycled. Different types of paper are usually sorted before recycling, such as newspapers and cardboard boxes.

Different grades of paper are recycled into different types of new products. Old newspapers are usually made into new newsprint, egg cartons, or paperboard. Old corrugated boxes are made into new corrugated boxes or paperboard. High-grade white office paper can be made into almost any new paper product: stationery, newsprint, magazines, or books.

Sometimes recyclers ask for the removal of the glossy inserts from newspapers because they are a different type of paper. Glossy inserts have a heavy clay coating that some paper mills cannot accept. Since the paper is weighed down by the clay coating, a paper mill gets more recyclable fibres from a ton of pure newsprint.

Paper can only be recycled a finite number of times due to the shortening of paper fibres making the material less versatile. Often it will be mixed with a quantity of virgin material, referred to as *downcycling*. This does not however exclude the material from being used in other processes such as composting or anaerobic digestion, where further value can be extracted from the material in the form of compost or biogas.

Plastic

Plastic recycling is the process of recovering scrap or waste plastics and reprocessing the material into useful products. Compared to glass or metallic materials, plastic poses unique challenges—because of the massive number of types of plastic, they each carry a resin identification code, and must be sorted before they can be recycled. This can be costly—while metals can be sorted using electromagnets, no such 'easy sorting' capability exists for plastics. In addition to this, while labels do not need to be removed from bottles for recycling, lids are often made from a different kind of non-recyclable plastic.

Plastics recycling rates lag far behind those of other items, such as newspaper and aluminium; consumers are typically unsure of how to recycle plastics, and compared to paper and metals fewer recycling facilities exist. Finally, recycled plastic is less appealing to manufacturers than new plastic.

Ship Breaking

A form of metal recovery associated to recycling is "ship breaking". This is the process of breaking a ship into smaller, recyclable pieces of metal. It often has a number of major drawbacks to the local community and the local environment where ship breaking occurs.

Ship breaking tends to occur in poor countries where lack of or insufficient safety standards, labour laws and wage agreements makes them a lucrative area for demolition work. India, Pakistan, Turkey and Bangladesh make up the majority of these countries.

Toxic material in the form of metals, gas, fumes and exhaust often contaminate a large area surrounding the ship breaking yards, including nearby villages and sleeping quarters for the workers, which are commonly located near the yards.

Material such as paint, electrical equipment, wire, anodes and coatings are often burned or simply dumped in the dismantling process. This releases metals such as mercury, lead, arsenic and chromium.

Polychlorinated organic compounds are another source of toxic material that can be found in transformers and cable insulation often burned or dumped in and around the ship breaking yard.

It is believed that many of the social, economical and environmental drawbacks in shipbreaking could be alleviated greatly by adhering to safe handling of the recycling process, or the ship owner decontaminating the toxins from the ship before it gets sent to be demolished.

Textiles

When considering textile recycling one must understand what the material consists of. Most textiles are composites of cotton (biodegradable material) and synthetic plastics. The textile's composition will affect its durability and method of recycling.

Workers sort and separate collected textiles into good quality clothing and shoes which can be reused or worn. These sorting facilities are in a trend of being moved from developed countries such as the UK to developing countries.

Damaged textiles are further sorted into grades to make industrial wiping cloths and for use in paper manufacture or material which is suitable for fibre reclamation and filling products. If textile reprocessors receive wet or soiled clothes however, these may still end up being disposed of in landfill, as the washing and drying facilities are not present at sorting units.

Fibre reclamation mills sort textiles according to fibre type and colour. Colour sorting eliminates the need to re-dye the recycled textiles. The textiles are shredded into "shoddy" fibres and blended with other selected fibres, depending on the intended end use of the recycled yarn. The blended mixture is carded to clean and mix the fibres and spun ready for weaving or knitting. The fibres can also be compressed for mattress production. Textiles sent to the flocking industry are shredded to make filling material for car insulation, roofing felts, loudspeaker cones, panel linings and furniture padding.

Timber

Recycling timber has become popular due to its image as an environmentally friendly product, with consumers commonly believing that by purchasing recycled wood the demand for *green timber* will fall and ultimately benefit the environment. Greenpeace also view recycled timber as an environmentally friendly product. The arrival of recycled timber as a construction product has been important in both raising industry and consumer awareness towards deforestation and promoting timber mills to adopt more environmentally friendly practices.

6

Wastewater Treatment

Industrial wastewater treatment covers the mechanisms and processes used to treat waters that have been contaminated in some way by man's industrial or commercial activities prior to its release into the environment or its reuse.

Most industries produce some wet waste although recent trends in the developed world have been to minimise such production or recycle such waste within the production process. However, many industries remain dependent on processes that produce a water based waste stream.

Industrial Wastewater

Agricultural Wastewater Treatment: Agricultural wastewater treatment relates to the treatment of wastewaters produced in the course of agricultural activities.

As agriculture is a highly intensified industry in many parts of the world, the range of wastewaters requiring treatment can encompass at least the following:

- Animals wastes - both liquid and solid
- Silage liquor

- Pesticide run off and surpluses
- Milking parlour wastes including milk
- Slaughtering waste
- Vegetable washing water
- Fire water

Animal Wastes

The constituents of animal wastewater typically contain:

- Strong organic content—much stronger than human sewage.
- High solids concentration.
- High nitrate and phosphorus content.
- Antibiotics.
- Synthetic hormones.
- Often high concentrations of parasites and their eggs.
- Spore of cryptosporidum—a bacterium resistant to drinking water treatment processes.
- Spore of Giardia.
- Human pathogenic bacteria such as Brucella and Salmonella.

Animal wastes from cattle can be as produced as solid or semi-solid manure or as a liquid slurry. The production of slurry is especially common in housed dairy cattle.

Treatment: Whilst solid manure heaps outdoors can give rise to polluting wastewaters from rain washing, this type of waste is usually relatively easy to treat by containment and/ or covering of the heap.

Animal slurries require special handling and are usually treated by containment in lagoons before disposal by spray or trickle application to grassland. Constructed wetlands are sometimes used to facilitate treatment of animal wastes, as are anaerobic lagoons. Excessive application or application to sodden land or insufficient land area can result in direct run-off to watercourses with the potential for causing severe pollution. Application of slurries to land overlying aquifers can result in direct contamination or, more commonly, elevation of nitrogen levels as nitrite or nitrate.

The disposal of any wastewater containing animal waste upstream of a drinking water intake can pose serious health problems to those drinking the water because of the highly resistant spores present in many animals that are capable of causing disabling in humans. This risk exists even for very low level seepage via shallow surface drains or from rainfall run-off.

Some animal slurries are treated by mixing with straws and composted at high temperature to produce a bacteriologically sterile and friable manure for soil improvement.

Piggery Waste

Piggery waste is comparable to other animal wastes except that many piggery wastes contain elevated levels of copper that can be toxic in the natural environment. Ascarid worms and their eggs are also common and can infect humans if wastewater treatment is ineffective.

Treatment: As for general animal waste although the liquid fraction of the waste is frequently separated off and reused in the piggery to avoid the prohibitively expensive costs of disposing of a copper rich liquor.

Silage Liquor

Fresh or wilted grass or other green crops can be made into the semi- fermented product called silage which can be stored and used as winter forage for cattle and sheep. The production of silage often involves the use of an acid conditioner such as sulphuric acid or formic acid. The process of silage making frequently produces a yellow-brown strongly smelling liquid which is very rich in simple sugars, alcohol, short-chain organic acids and silage conditioner. This liquor is one of the most polluting organic substances known. The volume of silage liquor produced is generally in proportion to the moisture content of the ensiled material.

Treatment: Silage liquor is best treated through prevention by wilting crops well before silage making. Any silage liquor that is produced can be used as part of the food for pigs. The most effective treatment is by containment in a slurry lagoon and subsequently spread on land following substantial dilution with slurry. Containment of silage liquor on its own can cause structural problems in concrete pits because of the acidic nature of silage liquor.

Pesticide Run-off and Surpluses

Inappropriate use of pesticides so that pesticide-containing wastewaters enter the environment can give rise to severe and long-lasting ecological damage. This is particularly true for insecticides used in sheep dips because of the volumes of pesticide-containing wastewater requiring disposal and because of the persistent and damaging nature of the pesticides.

Treatment: There are few safe ways of disposing of pesticide surpluses other than through containment in well managed landfills or by incineration. In some parts of the world, spraying on land is a permitted method of disposal.

Milking Parlour Wastes Including Milk

Although milk has a deserved reputation as an important and valuable food product, its presence in wastewaters is highly polluting because of its organic strength, which can lead to very rapid deoxygenation of receiving waters. Milking parlour wastes also contain large volumes of wash-down water, some animal waste together with cleaning and disinfection chemicals.

Treatment: Milking parlour waste are often treated in admixture with human sewage in a local sewage treatment plant. This ensures that disinfectants and cleaning agents are sufficiently diluted and amenable to treatment. Running milking wastewaters into a farm slurry lagoon is a possible option although this tends to consume lagoon capacity very quickly. Land spreading is also a treatment option.

Slaughtering Waste

Wastewater from slaughtering activities is similar to milking parlour waste although considerably stronger in its organic composition and therefore potentially much more polluting.

Treatment: As for milking parlour waste.

Vegetable Washing Water

Washing of vegetables produces large volumes of water contaminated by soil and vegetable pieces. Low levels of pesticides used to treat the vegetables may also be present together with moderate levels of disinfectants such as chlorine.

Treatment: Most vegetable washing waters are extensively recycled with the solids removed by settlement and filtration. The recovered soil can be returned to the land

Fire Water

Although few farms plan for fires, fires are nevertheless more common on farms than on many other industrial premises. Stores of pesticides, herbicides, fuel oil for farm machinery and fertilisers can all help promote fire and can all be present in environmentally lethal quantities in wastewater from firefighting at farms.

Treatment: All farm environmental management plans should allow for containment of substantial quantities of firewater and for its subsequent recovery and disposal by specialist disposal companies. The concentration and mixture of contaminants in fire-water make them unsuited to any treatment method available on the farm. Even land spreading has produced severe taste and odour problems for downstream water supply companies in the past.

Iron and Steel Industry

The production of iron from its ores involves powerful reduction reactions in blast furnaces. Cooling waters are inevitably contaminated with products especially ammonia, cyanide. Production of coke from coal in coking plants also requires water cooling and the use of water in by-products separation. Contamination of waste streams includes gasification products such as benzene, naphthalene, anthracene, cyanide, ammonia, phenols, cresols together with a range of more complex organic compounds known collectively as polyaromatic hydrocarbons (PAH).

The conversion of iron or steel into sheet, wire or rods requires hot and cold mechanical transformation stages frequently employing water as a lubricant and coolant. Contaminants include hydraulic oils, tallow and particulate solids. Final treatment of iron and steel products before onward sale into manufacturing includes pickling in strong mineral acid to remove rust and prepare the surface for tin or chromium plating or for other surface treatments such as galvanisation or painting. The two acids commonly used are hydrochloric acid and sulphuric acid. Wastewaters include acidic rinse waters together with waste acid. Although many plants operate acid recovery plants, (particularly those using Hydrochloric acid), where the mineral acid is boiled away from the iron salts, there remains a large volume of highly acid ferrous sulphate or ferrous chloride to be disposed of. Many steel industry wastewaters are contaminated by hydraulic oil also known as *soluble oil.*

Mines and Quarries

The principal wastewaters associated with mines and quarries are slurries of rock particles in water. These arise from rainfall washing exposed surfaces and haul roads and also from rock washing and grading processes. Volumes of water can be very high, especially rainfall related arisings on large sites. Some specialist separation operations such as coal washing to separate coal from native rock using density gradients can produce wastewater contaminated by fine particulate haematite and surfactants. Oils and hydraulic oils are also common contaminants. Wastewater from metal mines and ore recovery plants are inevitably contaminated by the minerals present in the native rock formations. Following crushing and extraction of the desirable materials, undesirable materials may become contaminated in the wastewater. For metal mines, this can include unwanted metals such as zinc and other materials such as arsenic. Extraction of high value metals such as gold and silver may generate slimes containing very fine particles in where physical removal of contaminants becomes particularly difficult.

Food Industry

Wastewater generated from agricultural and food operations has distinctive characteristics that set it apart from common municipal wastewater managed by public or private wastewater treatment plants throughout the world: it is biodegradable and non-toxic, but that has high concentrations of biochemical oxygen demand (BOD) and suspended solids (SS). The constituents of food and agriculture wastewater are often complex to predict due to the differences in BOD and pH in effluents from vegetable, fruit, and meat products and due to the seasonal nature of food processing and post-harvesting.

Processing of food from raw materials requires large volumes of high grade water. Vegetable washing generates waters with high loads of particulate matter and some dissolved organics. It may also contain surfactants.

Animal slaughter and processing produces very strong organic waste from body fluids, such as blood, and gut contents. This wastewater is frequently contaminated by significant levels of antibiotics and growth hormones from the animals and by a variety of pesticides used to control external parasites. Insecticide residues in fleeces is a particular problem in treating waters generated in wool processing.

Processing food for sale produces wastes generated from cooking which are often rich in plant organic material and may also contain salt, flavourings, colouring material and acids or alkali. Very significant quantities of oil or fats may also be present.

Organic Chemicals Industry

A range of industries manufacture or use complex organic chemicals. These include pesticides, Pharmaceuticals, paints and dyes, petrochemicals, detergents, plastics, etc. Waste waters can be contaminated by feed-stock materials, by-products, product material in soluble or particulate form, washing and cleaning agents, solvents and added value products such as plasticisers.

Water Treatment

Water treatment for the production of drinking water is dealt with elsewhere. Many industries have a need to treat water to obtain very high quality water for demanding purposes. Water treatment produces organic and mineral sludges from filtration and sedimentation. Ion exchange using natural or synthetic resins removes calcium, magnesium and carbonate ions from water, replacing them with hydrogen and hydroxyl ions. Regeneration of ion exchange columns with strong acids and alkalis produces a wastewater rich in hardness ions which are readily precipitated out, especially when in admixture with other wastewaters.

Treatment of Industrial Wastewater

The different types of contamination of wastewater require a variety of strategies to remove the contamination.

Solids Removal: Most solids can be removed using simple sedimentation techniques with the solids recovered as slurry or sludge. Very fine solids and solids with densities close to one pose special problems. In such case filtration or ultrafiltration may be required. Alternatively, flocculation may be used using alum salts or the addition of poly-electrolytes.

Oils and Greases: Many oils can be recovered from open water surfaces by skimming devices. However, hydraulic oils and the majority of oils that have degraded to any extent will also have a soluble or emulsified component that will require further treatment to eliminate. Dissolving or emulsifying oil using surfactants or solvents usually exacerbates the problem rather than solving it, producing a very difficult to treat wastewater.

Soft Organics: Organic material of plant or animal origin is usually possible to treat using extended conventional Wastewater treatment processes. Problems can arise if the wastewater is excessively diluted with washing water or is highly concentrated such as neat blood or milk. The presence of cleaning agents, disinfectants, pesticides, or antibiotics can have detrimental impacts on treatment processes.

Hard Organics: Synthetic organic materials including solvents, paints, pharmaceuticals, pesticides, coking products, etc. can be very difficult to treat. Treatment methods are often specific to the material being treated. Methods include distillation, adsorption, vitrification, incineration, chemical immobilisation or landfill disposal. Some materials such as some detergents may be capable of biological degradation and in such cases, a modified form of wastewater treatment can be used.

Acids and Alkalis: Acids and alkalis can usually be neutralised under controlled conditions. Neutralisation frequently produces a precipitate that will require treatment as a solid residue that may also be toxic. In some cases, gasses may be evolved requiring treatment for the gas stream. Some other forms of treatment are usually required following neutralisation.

Waste streams rich in hardness ions as from de-ionisation processes can readily loose the hardness ions in a buildup of precipitated calcium and magnesium salts. This precipitation process can cause severe *furring* of pipes and can, in extreme cases, cause the blockage of disposal pipes. A 1 metre diameter industrial marine discharge pipe serving a major chemicals complex was blocked by such salts in the 1970s. Treatment is by concentration of de-ionisation waste waters and disposal to landfill or by careful pH management of the released wastewater.

Toxic Materials

Toxic materials including many organic materials, metals (such as zinc, silver, cadmium, thallium, etc.) acids, alkalis, non-metallic elements (such as arsenic or selenium) are generally resistant to biological processes unless very dilute. Metals can often be precipitated out by changing the pH or by treatment with other chemicals. Many, however, are resistant to treatment or mitigation and may require concentration followed by landfilling or recycling.

Waste Treatment Technology

Anaerobic Clarigester: The anaerobic clarigester is a form of anaerobic digester. It is regarded as being the ancestor of the UASB anaerobic digester. A clarigester treats dilute biodegradable feedstocks and separates out solid and hydraulic (liquid) retention times.

Anaerobic Contact Process: The anaerobic contact process is a type of anaerobic digester. Here a set of reactors are created in series, often with recycling. This recycled material is pumped up into the bottom of the first reactor, an upflow reactor. The upflow anaerobic process is a large reactor which allows the waste to flow up from the bottom and separates the waste into 3 zones. At the very top is the biogas zone where the gas is collected. Bacteria digest waste in the lowest portion of the upflow reactor; the bioreactor zone. In between these two stages is the clarifier zone where the which exports the stabilised waste.

Anaerobic Digester Types: The following is a partial list of different types of anaerobic digesters. These processes and systems harness anaerobic digestion for purposes such as sewage treatment and biogas generation.

Examples Include:

- Anaerobic activated sludge process
- Anaerobic clarigester
- Anaerobic contact process
- Anaerobic expanded-bed reactor
- Anaerobic filter
- Anaerobic fluidised bed
- Anaerobic lagoon
- Anaerobic migrating blanket reactor
- Batch system anaerobic digester
- Expanded granular sludge bed digestion
- Hybrid reactor
- Imhoff tank
- One-stage anaerobic digester
- Submerged media anaerobic reactor
- Two-stage anaerobic digester
- Upflow anaerobic sludge blanket digestion
- Upflow and down-flow anaerobic attached growth.

Anaerobic Filter: An anaerobic filter (AF) is a form of anaerobic digester. The digestion tank contains a filter medium which anaerobic bacterial populations can establish upon. Anaerobic filters are commonly employed in the treatment of waste waters. AF reactors are gaining in popularity versus more established aerobic waste water treatment systems as they produce less solid residue.

AF reactors can be combined with UASB reactors. These reactors are called hybrid reactors.

Anaerobic Lagoon: Anaerobic lagoons are used to dispose of animal waste, particularly that of cows and pigs. The waste is washed into the lagoon by flushing the animal pens with water. Solid waste, particularly the fibrous type of cows, is sometimes separated before the wastewater enters the lagoon to prevent the buildup of solid material. Anaerobic organisms naturally present in the manure and the environment decompose the waste in the anaerobic conditions of the lagoon.

Areas with cold winters are inappropriate for anaerobic lagoons because the activity of the microorganisms is highly dependent on temperature. It is critical to have the proper size for the lagoon, with volume being more important than surface area. A minimum of two metres is necessary for anaerobic conditions, but the depth should not exceed 6 metres. Sometimes a secondary lagoon is used to accept wastes while the primary lagoon is undergoing maintenance or for other purposes.

If the anaerobic lagoon system is being used for energy production, the primary lagoon has a cover floating on the surface of the water. The cover captures the biogas produced by anaerobic bacteria. The biogas produced by anaerobic lagoons is 50 to 75 per cent methane, with carbon dioxide making up most of the rest. The gas is usually used to produce electricity using a microturbine or reciprocating engine, but it can also be used for water or space heating. The gas usually undergoes pretreatment, particularly dehydration, prior to combustion. Sometimes the carbon dioxide, which is incombustible, is also removed.

Enhanced Biological Phosphorus Removal: Enhanced biological phosphorus removal (EBPR) is a wastewater treatment configuration applied to activated sludge systems for the removal of phosphate.

The common element in EBPR implementations is the presence of an anaerobic tank (nitrate and oxygen are absent) prior to the aeration tank. Under these conditions a group of heterotrophic bacteria, called polyphosphate-accumulating organisms (PAO) are selectively enriched in the bacterial community within the activated sludge. These bacteria accumulate large quantities of polyphosphate within their cells and the removal of phosphorus is said to be enhanced.

Generally speaking, all bacteria contain a fraction (1-2 per cent) of phosphorus in their biomass due to the its presence in cellular components, such as membrane phospholipids and DNA. Therefore as bacteria in a wastewater treatment plant consume nutrients in the wastewater, they grow and phosphorus is incorporated into the bacterial biomass. When PAOs grow they not only consume phosphorus for cellar components but also accumulate large quantities of polyphosphate within their cells.

Thus, the phosphorus fraction of phosphorus accumulating biomass is 5-7 per cent. This biomass is then separated from the treated water at end of the process and the phosphorus is thus removed. Thus if PAOs are selectively enriched by the EBPR configuration, considerably more phosphorus is removed, compared to the relatively poor phosphorus removal in conventional activated sludge systems.

of phosphorus in their biomass due to the its presence in cellular components such as membrane phospholipids and DNA. Therefore as bacteria in a wastewater treatment plant consume nutrients in the wastewater, they grow and phosphorus is incorporated into the bacterial biomass. When PAOs grow they not only consume phosphorus for cellular components but also accumulate large quantities of polyphosphate within their cells.

Thus, the phosphorus fraction of phosphorus accumulating biomass is 5-7 per cent. This biomass is then separated from the treated water at end of the process and the phosphorus is thus removed. Thus if PAOs are selectively enriched by the EBPR configuration considerably more phosphorus is removed compared to the relatively poor phosphorus removal in conventional activated sludge systems.

7

Bacterial Metabolism

Water that has been used by people and is disposed into a receiving water body with altered physical and/or chemical parameters is defined as wastewater. If only the physical parameters of the water were changed, e.g., resulting in an elevated temperature after use as a coolant, treatment before final disposal into a surface water may require only cooling close to its initial temperature.

If the water, however, has been contaminated with soluble or insoluble organic or inorganic material, a combination of mechanical, chemical, and/or biological purification procedures may be required to protect the environment from periodic or permanent pollution or damage.

For this reason, legislation in industrialised and in many developing countries has reinforced environmental laws that regulate the maximum allowed residual concentrations of carbon, nitrogen, and phosphorous compounds in purified wastewater, before it is disposed into a river or into any other receiving water body.

However, enforcement of these laws is not always very

strict. Enforcement seems to be related to the economy of the country and thus differs significantly between wealthy industrialised and poor developing countries.

Decomposition of Organic Carbon Compounds

Catabolic processes of microorganisms, algae, yeasts, and lower fungi are the main pathways for total or at least partial mineralisation/decomposition of bioorganic and organic compounds in natural or man-made environments. Most of this material is derived directly or indirectly from recent plant or animal biomass. It originates from carbon dioxide fixation via photosynthesis ($\rightarrow$ plant biomass), from plants that served as animal feed ($\rightarrow$ detritus, faeces, urine, etc.), or from fossil fuels or biologically or geochemically transformed biomass ($\rightarrow$ peat, coal, oil, natural gas). Even the carbon portion of some xenobiotics can be tracked back to a biological origin, i.e., if these substances were produced from oil, natural gas, or coal. Only because the mineralisation of carbonaceous material from decaying plant and animal biomass in nature under anaerobic conditions with a shortage of water was incomplete, did the formation of fossil oil, natural gas, and coal deposits from biomass occur through biological and/or geochemical transformations. The fossil carbon of natural gas, coal, and oil enters the atmospheric CO_2 cycle again as soon as these compounds are incinerated as fuels or used for energy generation in industry or private households.

Biological degradation of recent biomass and of organic chemicals during solid waste or wastewater treatment proceeds either in the presence of molecular oxygen by respiration, under anoxic conditions by denitrification, or under anaerobic conditions by methanogenesis or sulphidogenesis. Respiration of soluble organic compounds or of extracellularly solubilised biopolymers such as carbohydrates, proteins, fats, or lipids in activated sludge systems leads to the formation of carbon dioxide, water, and a significant amount of surplus sludge. Some ammonia and H_2S may be formed during degradation of sulphur-containing amino acids or heterocyclic compounds. Oxygen must either be supplied by aeration or by injection of pure oxygen.

The two process variant for oxygen supply differ mainly in their capacity for oxygen transfer and the stripping efficiency for carbon dioxide from respiration.

Stripping of carbon dioxide is necessary to prevent a drop in pH and to remove heat energy. Respiration in the denitrification process with chemically bound oxygen supplied in the form of nitrate or nitrite abundantly yields dinitrogen. However, some nitrate escapes the reduction to dinitrogen in wastewater treatment plants and contributes about 2 per cent of the total N_2O emissions in Germany (Schon *et al.*, 1994).

Denitrifiers are aerobic organisms that switch their respiratory metabolism to the utilisation of nitrate or nitrite as terminal electron acceptors, if grown under anoxic conditions. Only if the nitrate in the bulk mass has been used completely does the redox potential become low enough for growth of strictly anaerobic organisms, such as methanogens or sulphate reducers. If anaerobic zones are allowed to form in sludge flocs of an activated sludge system, e.g., by limitation of the oxygen supply, methanogens and sulphate reducers may develop in the centre of sludge flocs and form traces of methane and hydrogen sulphide, found in the offgas.

Under strictly anaerobic conditions, soluble carbon compounds of wastes and wastewater are degraded stepwise to methane, CO_2, NH_3, and H_2S via a syntrophic interaction of fermentative and acetogenic bacteria with methanogens or sulphate reducers.

The complete methanogenic degradation of biopolymers or monomers via hydrolysis/fermentation, acetogenesis, and methanogenesis can proceed only at a low H_2 partial pressure, which is maintained mainly by interspecies hydrogen transfer.

Interspecies hydrogen transfer is facilitated when acetogens and hydrogenolytic methanogenic bacteria are arranged in proximity in flocs or in a biofilm within short diffusion distances. The reducing equivalents for carbon dioxide reduction to methane or sulphate reduction to sulphide are derived from the fermentative metabolism, e.g., of clostridia or *Eubacterium* sp., from a-oxidation of fatty acids, or the oxidation of alcohols. Methane and CO_2 are the main products in anaerobic environments where sulphate is absent, but sulphide and CO_2 are the main products if sulphate is present.

Basic Biology, Mass, and Energy Balance

To make soluble and insoluble biopolymers – mainly carbohydrates, proteins, and lipids – accessible for respiration by bacteria, the macromolecules must be hydrolysed by exoenzymes, which often are produced and excreted only after contact with respective inductors. The exoenzymes adsorb to the biopolymers and hydrolyse them to monomers or at least to oligomers. Only soluble, low molecular weight compounds (e.g., sugars, disaccharides, amino acids, oligopeptides, glycerol, fatty acids) can be taken up by microorganisms and be metabolised for energy production and cell multiplication.

Once taken up, degradation via glycolysis (sugars, disaccharides, glycerol), hydrolysis and deamination (amino acids, oligopeptides), or hydrolysis and a-oxidation (phospholipids, long-chain fatty acids) proceeds in the cells. Metabolism of almost all organic compounds leads to the formation of acetyl-CoA as the central intermediate, which is used for biosyntheses, excreted as acetate, or oxidised to CO_2 and reducing equivalents in the tricarboxylic acid (TCA) cycle. The reducing equivalents are respired with molecular oxygen in the respiration chain.

The energy of a maximum of only 2 mol of anhydridic phosphate bonds of ATP is conserved during glycolysis of 1 mol of glucose through substrate chain phosphorylation. An additional 2 mol of ATP are formed during oxidation of 2 mol of acetate in the TCA cycle, whereas 34 mol ATP are formed by electron transport chain phosphorylation with oxygen as the terminal electron acceptor. During oxygen respiration, reducing equivalents react with molecular oxygen in a controlled combustion reaction.

When carbohydrates are respired by aerobic bacteria, about one third of the initial energy content is lost as heat, and two thirds are conserved biochemically in 38 phosphoanhydride bonds of ATP. In activated sludge reactors or in wastewater treatment ponds that are not loaded with highly concentrated wastewater, wall irradiation and heat losses with the offgas stream of aeration into the atmosphere prevent self-heating. In activated sludge reactors for treatment of highly concentrated wastewater, however, self-heating up to thermophilic temperatures may occur if the wastewater is warm in the beginning, the hydraulic retention time for biological treatment is short (short aeration time), and the air or oxygen stream for aeration is restricted so as to supply just sufficient oxygen for complete oxidation of the pollutants (small aeration volume).

The conserved energy in the terminal phosphoanhydride bond of ATP, formed during substrate chain and oxidative phosphorylation by proliferating bacteria is partially used for maintenance metabolism and partially for cell multiplication. Partitioning between both is not constant, but depends on the nutritional state. In highly loaded activated sludge reactors with a surplus or at least a non-growth–limiting substrate supply, approximately 50 per cent of the substrate is respired in the energy metabolism of the cells and 50 per cent serves as a carbon source for cell growth.

The biochemically conserved energy must be dissipated to be used for the maintenance metabolism of existing cells and cell growth.

If the substrate supply is growth-limiting, e.g., in a low-loaded aerobic treatment system, a higher proportion of ATP is consumed for maintenance, representing the energy proportion that bacteria must spend for non-growth–associated cell survival metabolism, and less energy is available for growth.

Overall, more of the substrate carbon is respired, and the ratio of respiration products to surplus sludge formed is higher, e.g., around 70 per cent: 30 per cent. In a trickling filter system, an even higher proportion of the substrate seems to be respired. This might be due to protozoa grazing off part of the biofilm.

Mass and Energy Balance for Aerobic Glucose Respiration and Sewage Sludge Stabilisation

Except for an exact reaction stoichiometry of the oxidative metabolism, mass and energy dissipation, if mentioned at all, are not quantified. Both parameters are, however, very important for activated sludge treatment plants. The surplus sludge formed during wastewater stabilisation requires further treatment, causes disposal costs, and – in the long run – may be an environmental risk, and heat evolution during unevenly high-loaded aerobic treatment may shift the population towards more thermotolerant or thermophilic species and thus, at least for some time, may decrease the process efficiency.

$$1 \text{ mol } C_6H_{12}O_6 + 6 \text{ mol } O_2 \rightarrow 6 \text{ mol } CO_2 + 6 \text{ mol } H_2O + \text{heat energy} \quad (1)$$

Table: Carbon flow during (A) aerobic degradation in an activated sludge system under (a) saturating and (b) limiting substrate supplya and during (B) anaerobic degradation.

(A) Aerobic degradation:

(a) Saturating substrate supply = high-load conditions.

1 unit substrate carbon → 0.5 units CO_2 carbon + 0.5 units cell carbon.

(b) Limiting substrate supply = low-load conditions 1 unit substrate carbon → 0.7 units CO_2 carbon + 0.3 units cell carbon.

(B) Anaerobic degradation:

1 unit substrate carbon → 0.95 units (CO_2 + CH_4) carbon + 0.05 unit cell carbon.

If 1 mol of glucose (MW=180 g) is degraded in an activated sludge system at a high BOD loading rate (e.g., >0.6 kg m^{-3} d–1 BOD), approximately 0.5 mol (90 g) is respired to CO_2 and water, with consumption of 3 mol of O_2 (96 g), releasing 19 mol of ATP. The other 0.5 mol of glucose (90 g) is converted to pyruvate by one of three glycolytic pathways, accompanied by the formation of 0.5-1 mol ATP. Pyruvate or its subsequent metabolic products, e.g., acetate or dicarboxylic acids, are directly used as carbon substrates for cell multiplication and surplus biomass formation.

A maximum amount of 20 mol ATP is thus available for growth and maintenance. At a pH of 7, about 44 kJ of energy is available for growth per mol of ATP hydrolysed to ADP and inorganic phosphate (Thauer *et al.*, 1977). For an average molar growth yield of aerobes of 4.75 g per mol ATP (Lui, 1998), 90 g biomass can be generated from 180 g glucose. If the combustion energy per g of cell dry mass is 22 kJ, about 890 kJ (2870-980 kJ) is lost as heat during respiration.

The energy loss is the sum of heat losses during respiration and cell growth. At a low BOD loading rate, the proportion of glucose respired in relation to the proportion of glucose fixed as surplus biomass can shift. Up to 0.7 mol (126 g) of glucose can be oxidised to CO_2, requiring 4.2 mol of oxygen (134.4 g O_2). Thus, for respiration of 1 mol of glucose, different amounts of oxygen may be consumed, depending on the loading rate of the wastewater treatment system and the different amounts of carbon dioxide and of surplus sludge formed.

The energy and carbon balance deduced above can be analogously applied to aerobic stabilisation of raw sewage sludge. If the initial dry matter content is around 36g L^{-1} (average organic dry matter content of sewage sludge) and if a biodegradability of 50 per cent within the residence time in the sludge reactor is obtained, about 9 g L^{-1} of new biomass is formed, and thus 27 g L^{-1} (36 – 18 + 9) remains in the effluent.

The released heat energy is approximately 89 kJ L^{-1} of reactor content. To estimate the theoretical temperature rise, this amount of heat energy must be divided by 4.185 kJ (specific energy requirement for heating 1 L of H_2O from 14.5–15.5°C).

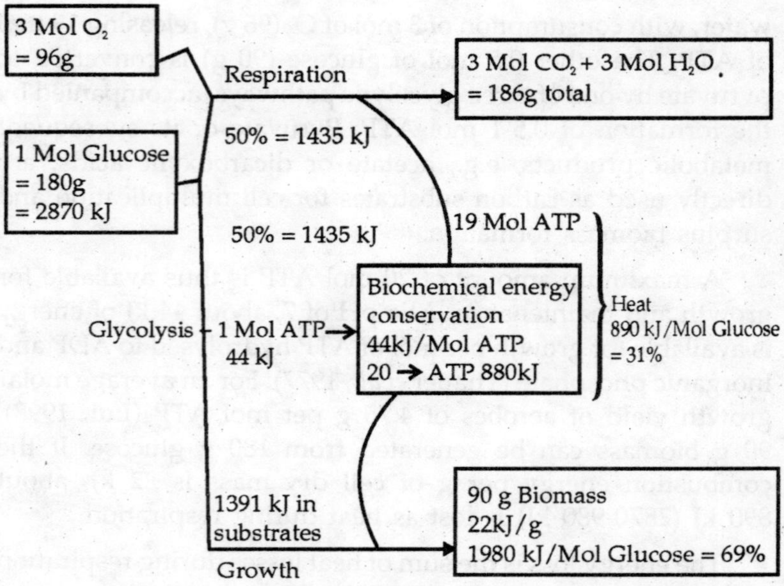

Mass and energy dissipation during glocose respiration at pH 7

Thus, by respiration of 18 g L^{-1} organic dry matter, the reactor temperature increases by 21.3°C within the residence time required for degradation ($\leq$ 16 h), provided that no heat energy is lost. A great proportion of the heat energy is, however, transferred via the liquid phase to the aeration gas and stripped out, whereas a smaller proportion is lost through irradiation from the reactor walls. Since air, containing almost 80 per cent nitrogen, is normally used as an oxygen source in aeration ponds or activated sludge reactors, the heat transfer capacity of the offgas is high enough to prevent a significant increase in the wastewater temperature.

Thus, ambient or at least mesophilic temperatures can be maintained. An increasing temperature of several degrees Celsius would lead to a shift in the population in the reactor and – at least temporarily – would result in reduced process stability, but an only slightly increased temperature of a few degrees Celsius might simply stimulate the metabolic activity of the prevalent mesophilic population. In practice, in activated sewage sludge systems no self-heating is observed because degradability is only about 50 per cent and complete heat transfer to the atmosphere occurs via the offgas at a retention time of more than 0.5 d. If, however, wastewater from a dairy plant or a brewery with a similar COD concentration, but with almost 100 per cent biodegradable constituents, is stabilised with pure oxygen, twice as much heat evolves, leading to a theoretical temperature rise of 57°C. Self-heating is observed, since there is much less offgas and the heat loss is thus significantly lower. In addition, due to higher reaction rates than with sewage sludge, the heat is generated during a shorter time span (shorter retention time).

Mass and Energy Balance for Anaerobic Glucose Degradation and Sewage Sludge Stabilisation

For anaerobic wastewater or sludge treatment, oxygen must be excluded to maintain the low redox potential that is required for survival and metabolic activity of the acetogenic, sulphidogenic, and methanogenic populations. Hydrolysis of polymers, uptake of soluble or solubilised carbon sources, and the primary metabolic reactions of glycolysis up to pyruvate and acetate formation seem to proceed identically or at least analogously in aerobic and anaerobic bacteria. Whereas aerobes oxidise acetate in the TCA cycle and respire the reducing equivalents with oxygen, anaerobes, such as *Ruminococcus* sp., *Clostridium* sp., or *Eubacterium* sp., either release molecular hydrogen or transform pyruvate or acetate to highly reduced metabolites, such as lactate, succinate, ethanol, propionate, or *n*-butyrate. For further degradation within the anaerobic food chain, these reduced metabolites must be oxidised anaerobically by acetogenic bacteria. Since the anaerobic oxidation of propionate or *n*-butyrate by acetogenic bacteria is obligately accompanied by hydrogen production but is only slightly exergonic under conditions of a low H_2 partial pressure (Bryant, 1979), acetogens can grow only when hydrogen is consumed by hydrogen-scavenging organisms such as methanogens or sulphate reducers.

During anaerobic degradation of 1 mol glucose, approximately 95 per cent of the glucose carbon is used for biogas formation (171 g = 127.7 L CH_4 + CO_2), and only about 5 per cent of the substrate carbon (9 g) converted to biomass. Much less heat energy is released during anaerobic metabolism than during aerobic respiration (131 kJ mol^{-1} versus 890 kJ mol^{-1}, respectively), and the biogas contains almost 90 per cent of the energy of the fermented substrate. Due to the heat energy requirement to warm the wastewater and due to heat losses via irradiation from pipes and reactor walls, heat generation is not nearly sufficient to maintain a constant mesophilic fermentation temperature. For this reason, anaerobic digesters must be heated.

In sewage sludge with 36 g L^{-1} organic dry matter content and 50 per cent biodegradability, 0.9 g surplus sludge and 17.1 g biogas (equivalent to 12.75 L) are formed during anaerobic stabilisation. Only 13.1 kJ of heat energy per mol of glucose is released, leading to a self-heating potential of 3.1°C. Since the heat energy is released only during the hydraulic residence time of the wastewater in the reactor, which is usually more than 10 d (except for high-rate industrial wastewater treatment, where in special situations the HRT may be shorter than one day), much more heat energy is lost by irradiation via the reactor walls than is required to maintain the temperature. If highly concentrated wastewater streams of the food and beverage industry are stabilised anaerobically at hydraulic retention times of <1 d (high space loading), more heat energy is generated within a much shorter time. But even then, process energy from external sources has to be supplied to maintain a temperature of 30-37°C. Proper insulation of anaerobic reactors can minimise but not replace external heating of the reactor. The methane in the biogas, generated during anaerobic sludge or wastewater stabilisation processes, contains about 90 per cent of the energy of the fermented substrate. Since methane is a climate-relevant gas it may not be emitted into the atmosphere but must be combusted to CO_2. Methane from anaerobic reactors can be used as a fuel for gas engines to generate electricity and/or heat energy (Eq. 2):

$$CH_4 + 2\ O_2 \rightarrow CO_2 + 2\ H_2O + \text{heat energy } (\Delta G^{\circ\prime} = -\ 891.6 \text{ kJ mol}^{-1}\ CH_4) \quad (2)$$

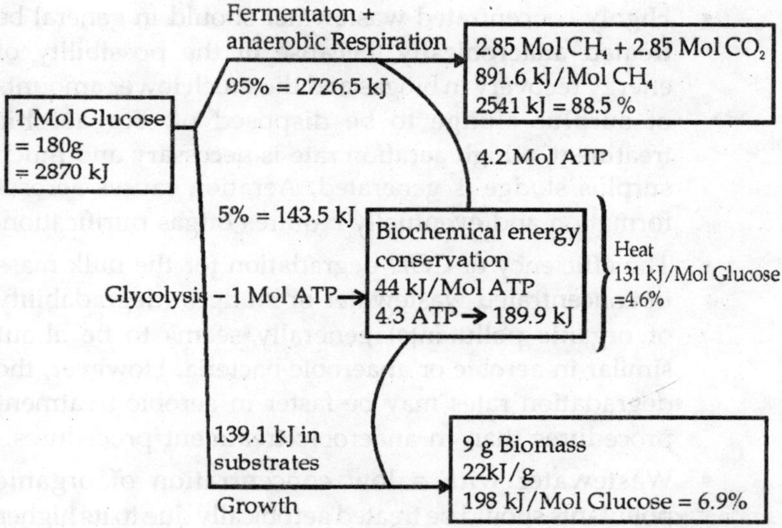

Mass and energy dissipation during anaerobic glucose fermentation at pH 7

General Considerations for the Choice of Aerobic or Anaerobic Wastewater Treatment Systems

If a producer of wastewater has to decide whether to install an aerobic or an anaerobic waste or wastewater treatment system, several points should be considered:

- Anaerobic treatment in general does not lead to the low pollution standards of COD, BOD5, or TOC that can be met with aerobic systems and which are required by environmental laws. Anaerobic treatment of wastes and wastewater is often considered a pretreatment process to minimise the oxygen demand and surplus sludge formation in a subsequent aerobic post-treatment stage. Only after a final aerobic treatment can the COD, BOD5, or TOC concentration limits stated in the environmental laws be met. If limiting concentrations for nitrogen and phosphate also have to be achieved, further treatment steps such as nitrification, denitrification, and biological or chemical phosphate removal, must be considered.

- Highly concentrated wastewater should in general be treated anaerobically, because of the possibility of energy recovery in biogas and the much lower amounts of surplus sludge to be disposed of. For aerobic treatment, a high aeration rate is necessary and much surplus sludge is generated. Aeration causes aerosol formation and eventually requires offgas purification.
- The efficiency of COD degradation for the bulk mass in concentrated wastewater or sludges (degradability of organic pollutants) generally seems to be about similar in aerobic or anaerobic bacteria. However, the degradation rates may be faster in aerobic treatment procedures than in anaerobic treatment procedures.
- Wastewater with a low concentration of organic pollutants should be treated aerobically due to its higher process stability at low pollutant concentrations, although aerobic treatment is more expensive and more sludge remains for disposal. If mineralised sludge is required, aerobic treatment at a low loading or at prolonged hydraulic retention times is necessary to reinforce respiration of all endogenous reserve material.
- Anaerobic treatment systems are more expensive to construct but less expensive to operate than aerobic treatment systems.

Aerobic or Anaerobic Hydrolysis of Biopolymers

Kinetic Aspects: Hydrolysis of biopolymers and fermentation or respiration of monomers can be catalysed by strictly anaerobic, facultative anaerobic, and aerobic microorganisms.

With some exceptions (e.g., small protein molecules,

dextran), biopolymers are insoluble and form fibres (cellulose), grains (starch) or globules (casein after enzymatic precipitation) or can be melted or emulsified (fat). Henze *et al.* (1997) reported hydrolysis constants *k*h for dissolved organic polymers of 3–20 d^{-1} under aerobic conditions and of 2–20 d–1 under anaerobic conditions, whereas for suspended solids the hydrolysis constants *k*h were 0.6 to 1.4 d^{-1} under aerobic conditions and 0.3 to 0.7 d^{-1} under anaerobic conditions. For a kinetic description of hydrolysis and fermentation, a substrate-limited first-order reaction was assumed by Buchauer (1997), who deduced that the temperature-dependent reaction rate for hydrolysis is a little lower than the reaction rate for fermentation of the hydrolysis products. The ratelimiting step is therefore hydrolysis of particles and not fermentation of solubilised material (Buchauer, 1997). Since hydrolysis is catalysed not only by freely soluble exoenzymes, diluted in the bulk mass of liquid, but to a much higher extent by enzymes that are excreted in the neighbourhood of bacterial colonies growing attached to the surface of the particles, the above description of complex fermentation processes is not always valid. Cellulases can be arranged in cellulosomes, which attach to the particles, which in turn serve as carriers until they themselves are solubilised.

For this reason, Vavilin *et al.* (1997) included biomass in their description of the hydrolysis of cellulose, cattle manure, and sludge. Shin and Song (1995) determined the maximum rates of acidification and methanation for several substrates. For hydrolysis of particulate organic matter, the ratio of surface area to particle size is important.

They found that for glucose, starch, carboxymethyl cellulose, casein, and food residues from a restaurant, hydrolysis proceeded faster than methanogenesis, whereas for newspaper and leaves hydrolysis was the rate-limiting step.

Hydrolysis of Cellulose by Aerobic and Anaerobic Microorganisms

Biological Aspects: Cellulose and lignin are the main

structural compounds of plants. Both substances are the most abundant biopolymers on earth. Cellulose fibres are formed of linear chains of 100–1400 glucose units linked together by β-1,4-glycosidic bonds. Interand intramolecular hydrogen bonds and van der Waals interactions arrange the highly organised fibrous regions (crystalline region), which alternate with less organised amorphous regions in the cellulose fibres. The fibres are embedded in a matrix of hemicelluloses, pectin, or lignin. The hemicelluloses consist mainly of xylans or glucomannans, which have sidechains of acetyl, gluconuryl, or arabinofuranosyl units.

To make cellulose fibres accessible to microorganisms, the hemicellulose, pectin, or lignin matrix must be degraded microbiologically or solubilised chemically. Cellulose degradation in the presence of oxygen in soil or in the absence of oxygen in the rumen of ruminants, in swamps, or in anaerobic digesters is the most important step in mineralisation of decaying plant material. Cellulolytic organisms are found among aerobic soil fungi, e.g., within the genera *Trichoderma* and *Phanaerochaete* and in anaerobic rumen fungi, e.g. *Neocallimastix* and *Piromyces*, and among bacteria, e.g., within the genera *Cellulomonas, Pseudomonas,* and *Thermomonospora* (aerobic cellulose degraders) and *Clostridium, Fibrobacter, Bacteroides,* and *Ruminococcus* (anaerobic cellulose degraders). For more details on cellulolytic bacteria and the mechanism of cellulose cleavages.

Glycosyl hydrolases are involved in cellulose and hemicellulose degradation by cleaving glycosidic bonds between different carbohydrates and between carbohydrates and non-carbohydrates. Endo - and exocellulases – in some organisms organised in cellulosomes – must be excreted into the medium. Cellulases are complex biocatalysts and contain a catalytic site and a substrate-binding site. The presence of a non-catalytic substrate binding site permits tight attachment to the different forms of cellulose substrate and keeps the enzyme close to its cleaving sites. Substrate binding is reversible, which allows the enzyme to 'hike' along the fibres and obtain total solubilisation. Many aerobic fungi and some bacteria excrete endoglucanases that hydrolyse the amorphous region of cellulose (degradation within the chain), whereas exoglucanases hydrolyse cellulose from the ends of the glucose chains. Cellobiose is cleaved off by cellobiohydrolases from the non-reducing ends in the amorphous region, and finally the crystalline region is also hydrolysed.

Cleavage of cellobiose to glucose units by a-glucosidases is necessary to prevent cellobiose accumulation, which inhibits cellobiohydrolases (Be'guin and Aubert, 1994).

Anaerobic bacteria such as *Clostridium thermocellum* form a stable enzyme complex, the cellulosome, at the cell surface (Lamed and Bayer, 1988). Cellulosomes are active in degrading crystalline cellulose. The catalytic subunits of a cellulosome, endoglucanases and xylanases, cleave the cellulose fibre into fragments, which are simultaneously degraded further by a-glucosidases. Cellulosome-like proteins are found also in *Ruminococcus* sp. and *Fibrobacter* sp. cultures. The cell-bound enzymes are associated with the capsule or the outer membrane. Other specific adhesions or ligand formations with the cellulose can be facilitated by fimbrial connections, glycosylated epitopes of carbohydrate binding proteins, or the glycocalyx and carbohydrate binding modules (Krause *et al.*, 2003). Some bacteria have not developed a mechanism to adhere to cellulose fibres, but excrete cellulases into the medium.

Adsorption of bacteria onto cellulose fibres via cellulosomes offers the advantage of close contact with the substrate, which is hydrolysed mainly to glucose, which is then taken up and metabolised. Small amounts of cellobiose must be present initially to induce cellulase expression. The contact of bacteria with the solid substrate surface keeps them close to cellobiose and thus keeps cellulase activities high. However, accumulation of hydrolysis products such as glucose repress cellulase activity.

In nature most cellulose is degraded aerobically. Only about

5 per cent-10 per cent is thought to be degraded anaerobically, which may be an underestimate. Since most ecosystems are rich in carbonaceous substances but deficient in nitrogen compounds, many cellulolytic bacteria can also fix dinitrogen. This is advantageous to them and to syntrophic or symbiotic organisms (Leschine, 1995). Other examples of mutual interactions between organisms of an ecosystem are interspecies hydrogen transfer between anaerobes (anaerobic food chain), transfer of growth factors (mycorhizzae, Kefir), and production of fermentable substrates for the partner organisms (bacterial interactions in the rumen).

Hydrolysis of biological structural components such as cellulose, lignin, and other structural or storage polymers is difficult. The limiting step of hydrolysis seems to be liberation of the cleavage products. In contrast to the slow hydrolysis of celluloses, mainly due to lignin encrustation of naturally occurring celluloses, starch can be easily hydrolysed. The branching and helical structure of starch facilitates hydrolysis (Warren, 1996). Whereas cellulose forms fibres with a large surface covered with lignin, starch forms grains with an unfavourable surface-to-volume ratio for enzymatic cleavage. Thus, although amylases may be present in high concentrations, the hydrolysis rate is limited by the limited access of the enzymes to the substrate.

Whereas cellulose and starch are biodegradable, other carbohydrate-derived cellular compounds are not biodegradable and – after reaction with proteins – form humic acid-like residues by the Maillard reaction.

Polysaccharides an d derivatives occuring in nature

Compound	*Bond*	*Unit*	*Occurrence*
Cellulose	β -1,4-	glucose	plant cell wall
Chitin	β -1,4-	*N*-acetyl glucosamine	fungal cell and insect wall exoskeleton
Murein	β -1,4-	*N*-acetyl-glucosamine *N*-acetyl-muramic acid	bacterial cell wall

Contd...

Compound	*Bond*	*Unit*	*Occurrence*
Chitosan	β -1,4-	*N*-glucosamine, substituted with acetyl residues	fungal and insect materials
Mannans	β -1,4-	mannose	plant material
Xylans	β -1,4-	xylopyranose, substituted with acetyl-arabinofuranoside residues	plant material
Starch	α -1,4-	glucose amylose (contains few α -1,6-branches) amylopectin (contains many α -1,6-branches)	storage material in plant
Glycogen	α -1,4-	highly branched glucose with α -1,6 bonds	storage material in bacteria
Dextran	α -1,2-	glucose	storage material in yeasts
	α -1,3-	glucose	exopolymer of bacteria
	α -1,4-	glucose	
	α -1,6-	glucose	
Laminarin	β -1,3-	glucose	reserve material in algae

Biomass Degradation in the Presence of Inorganic Electron Acceptors and by an Anaerobic Food Chain

In ecosystems in which molecular oxygen is available, plant and animal biomass is degraded to CO_2 and H_2O, catalysed by either single species of aerobic microorganisms or the whole population of the ecosystem, in competition for the substrates. A single organism may be able to hydrolyse the polymers and oxidise the monomers to CO_2 and H_2O with oxygen.

In ecosystems where molecular oxygen is deficient – such as swamps, wet soil, the rumen of animals, the digestive tract of humans, or in river and lake sediments – oxidation of dead biomass proceeds anoxically by reduction of electron acceptors such as nitrate and nitrite or anaerobically by reduction of sulphate, Fe^{3+}, Mn^{4+}, or CO_2. The oxidation of the carbon source is either complete or incomplete with acetate excretion. In the absence of inorganic electron acceptors, oxidised metabolites such as pyruvate or acetate are reduced to lactate or ethanol or biotransformed to, e.g., *n*-butyrate or *n*-butanol. In permanently anaerobic ecosystems with seasonal overfeeding, periodic accumulation of such metabolites can occur, e.g., in autumn after the non-evergreen plants drop their leaves or decay completely.

The biopolymers of the leaves or the plants themselves decompose by extracellular enzymatic hydrolysis. The monomers are fermented, and the fermentation products may be degraded further to biogas by acetogenic and methanogenic bacteria.

Whereas single cultures of aerobes can catalyse the whole mineralisation process to finally form CO_2 and H_2O, single cultures of strictly anaerobic bacteria are not capable of degrading biopolymers to CH_4 and CO_2. Under anaerobic conditions biopolymers must be degraded by a food chain via depolymerisation (hydrolysis), fermentation (acidogenesis), oxidation of fatty acids (acetogenesis), and biogas formation (methanogenesis) as the last step (McInerney, 1988). In an initial exoenzyme-catalysed reaction the biopolymers are hydrolysed to soluble mono-, di-, or oligomers.

These are taken up by the bacteria and fermented to CO_2, H_2, formate, acetate, propionate, butyrate, lactate, etc. If fatty acid isomers are produced, they are mainly derived from degradation of amino acids after proteolysis. Fatty acids are further oxidised by acetogenic bacteria, before the cleavage products CO_2, H_2, and acetate can be taken up by methanogens and be converted to methane and CO_2. Lactate is oxidised to pyruvate, which is decarboxylated to yield acetate, CO_2, and H_2. If ethanol is present, it is oxidised to acetate and hydrogen, and the hydrogen is used for CO_2 reduction.

Summarises the reactions that can be catalysed by methanogens and that can contribute to methane emission in various ecosystems. In sewage digesters about two thirds of the methane is derived from acetate cleavage and one third from CO_2 reduction with H_2. If hexoses are the substrates and glycolysis is the main degradation pathway, then the 2 mol of pyruvate can be decarboxylated by pyruvate: ferredoxin oxidoreductase to yield 2 mol acetate and 2 mol CO_2. The hydrogens of the 2 mol $NADH_2$ from glycolysis and the 2 mol FdH_2 from pyruvate decarboxylation are then released as molecular hydrogen at low H_2 partial pressure (Eq. 3). Two moles of CH_4 are then formed from acetate and 1 mol of CH_4 by CO_2 reduction.

1 mol glucose → 2 mol acetate + 2 mol CO_2 + 4 mol H_2 (at low *p*H2) (3)

Reactions catalyzed by methanogens and standard changes in free energy

Reaction: Substrates (mol)		Products (mol)	Δ G°′ (kJ per mol of methane)
Acetate	→	$CH_4 + CO_2$	–31.0
$4\ H_2 + CO_2$	→	$CH_4 + 2\ H_2O$	–135.6
4 HCOOH	→	$CH_4 + 3\ CO_2 + 2\ H_2O$	–130.1
$4\ CO + 2\ H_2O$	→	$CH_4 \rightarrow 3\ CO_2$	–211.0
4 Methanol	→	$3\ CH_4 + CO_2 + 2\ H_2O$	–104.9
Methanol + H_2	→	$CH_4 + H_2O$	–112.5
2 Ethanol[a] + CO_2	→	CH_4 + 2 acetate	–116.3
4 2-Propanol[b] + CO_2	→	CH_4 + 4 acetone + 2 H_2O	–36.5
4 Methylamine + 2 H_2O	→	$3\ CH_4 + CO_2 + 4\ NH_3$	–75.0
2 Dimethylamine + 2 H_2O	→	$3CH_4 + CO_2 + 2\ NH_3$	–73.2
4 Trimethylamine + 6 H_2O	→	$9\ CH_4 + 3\ CO_2 + 4\ NH_3$	–74.3
2 Dimenthysulfide + 2 H_2O	→	$3\ CH_4 + CO_2 + 2\ H_2S$	–73.8

In complex ecosystems formate is formed if high concentrations of hydrogen accumulate.

Syntrophic interactions are usually associated with interspecies hydrogen transfer, but evidence for interspecies formate transfer was also reported (Thiele *et al.*, 1988). The feasibility of the electron carrier depends on its solubility, which is much less for hydrogen than for formate, and on its diffusion coefficient in water, which favours hydrogen 30 times over formate. The efficiency of the appropriate electron transfer depends mainly on the distance between the producing and consuming bacteria.

It can be expected, that formate transfer is favoured when the distance between communicating bacteria is high and hydrogen transfer when the distance is small (de Bok *et al.*, 2004). Interspecies formate transfer is thought to play a major role in degradation of syntrophic butyrate (Boone *et al.*, 1989) and propionate (Stams, 1994; Schink, 1997).

However, at an increased H_2 partial pressure formate is also produced by methanogens, either in pure cultures or in a sewage sludge population (Bleicher and Winter, 1994), and this may also contribute to increasing formate concentrations.

Other substrates for methanogenic bacteria, such as methanol (derived, e.g., from methoxy groups of lignin monomers) or methyl-amines and dimethylsulphide (e.g., from methylsulphonopropionate in algae; Fritsche, 1998) are relevant only in ecosystems where these substances are produced during microbial decay. A few methanogens can also use reduced products such as primary, secondary, and cyclic alcohols as a source of electrons for CO_2 reduction (Widdel, 1986; Zellner and Winter, 1987a; Bleicher *et al.*, 1989; Zellner *et al.*, 1989).

Roles of Molecular Hydrogen and Acetate During Anaerobic Biopolymer Degradation

Molecular hydrogen is produced during different stages of anaerobic degradation.

In the fermentative stage, organisms such as *Clostridium* sp. and *Eubacterium* sp. produce fatty acids, CO_2, and hydrogen from carbohydrates. In the acetogenic stage, acetogens such as *Syntrophobacter wolinii* and *Syntrophomonas wolfei* produce acetate, CO_2, and hydrogen or acetate and hydrogen by anaerobic oxidation of propionate and *n*-butyrate (McInerney, 1988).

Fermentative bacteria release molecular hydrogen even at a high H_2 partial pressure and simultaneously excrete reduced products (e.g., clostridia, *Ruminococcus, Eubacterium* sp.). However, the release of molecular hydrogen during acetogenesis of fatty acids or of other reduced metabolites may occur only when hydrogen does not accumulate, for thermodynamic reasons.

Molecular hydrogen is consumed by methanogens or, alternatively, by sulphate reducers via interspecies hydrogen transfer.

In the rumen and in sewage sludge digesters, the hydrogen concentration can be decreased by acetate formation from CO_2 and H_2 by bacteria such as *Acetobacterium woodii* and *Clostridium thermoaceticum*.

To maintain a low H_2 partial pressure, a syntrophism of acetogenic, hydrogenproducing and methanogenic, hydrogen-utilising bacteria is essential (Ianotti *et al.*, 1973). Complete anaerobic degradation of fatty acids with hydrogen formation by obligate proton-reducing acetogenic bacteria is possible only at H_2 partial pressures $<10^{-4}$ atm (*n*-butyrate) or 10^{-5} atm (propionate), which cannot be maintained by methanogens or sulphate reducers. However, by reversed electron transport electrons can be shifted to a lower redox potential suitable for proton reduction (Schink, 1997).

If hydrogen accumulates beyond this threshold concentration, the anaerobic oxidation of fatty acids becomes endergonic and does not proceed. Whereas hydrogen prevents β-oxidation of fatty acids by acetogens even at very low H_2 partial pressure, much higher concentrations of acetate (in the millimolar range) are required for the same effect.

Hydrogen-consuming reactions in anaerobic ecosystems (Schink, 1997)

Substrates (mol)		*Products (mol)*	$\Delta G^{o\prime}$ (kJ per mol)
$4 H_2 + CO_2$	→	$CH_4 + 2 H_2O$	-131.0
$4 H_2 + SO_4^{2-}$	→	$S^{2-} + 4 H_2O$	-151.0
$4 H_2 + 2 CO_2$	→	$CH_3COO^- + H^+ + 2 H_2O$	-0.9
$H_2 + S^0$	→	H_2S	-0.9
$H_2C(NH_3^+)COO^- + H_2$	→	$CH_3COO^- + NH_4^+$	0.0
$COOH\text{-}CH\text{-}CH\text{-}COOH + H_2$	→	$COOH\text{-}CH_2\text{-}CH_2\text{-}COOH$	0.0

The fermentative metabolism of acidogenic bacteria is exergonic even at H_2 partial pressures $>10^{-4}$ atm. Whereas acetogenic bacteria apparently depend mainly on ATP generation by chemiosmotic phosphorylation, fermentative bacteria produce most of their ATP by substrate chain phosphorylation. This may be why fermentative bacteria do not depend on a syntrophic interaction with electron-consuming bacteria, such as methanogens or sulphate reducers.

In addition to the possibility of anaerobic oxidation of organic compounds via synthrophic interactions between acetogenic bacteria and acetoclastic + hydrogenotrophic methanogens, other synthrophic associations between acetate-oxidising bacteria and H_2/CO_2-utilising methanogens under thermophilic (Lee and Zinder 1988) and mesophilic (Schnuerer *et al.* 1996) growth conditions have been observed. Thermodynamic analysis of mesophilic synthrophic acetate oxidation revealed a hydrogen partial pressure of < 0.1–2.6 Pa, which is in the range found in methanogenic ecosystems (Dolfing, 2001). In thermophilic methanogenic reactors, acetate is degraded either by synthrophic acetate oxidisers (dominant process at low acetate concentrations) and acetate-degrading methanogens (acetate concentration above the threshold concentration) or by acetate-utilising methanogens of the genera *Methanosaeta* or *Methanosarcina* (Ahring, 2003). Synthrophic acetate oxidation and methane formation from the cleavage products may explain the lack of acetoclastic methanogens (*Methanosarcina* sp. or *Methanosaeta* sp.) in anaerobic reactors.

Anaerobic Conversion of Biopolymers to Methane and CO_2

The principle of the anaerobic metabolism of biopolymers was outlined by the pioneering work of Wolin (1976) and Bryant (1979). There is an essential requirement for syntrophic interaction between different metabolic groups for complete anaerobic degradation (Wolin, 1976, 1982). The most sensitive switch of the carbon flow of substrates to biogas is the H_2 partial pressure. The substrate supply for biomethanation processes must be limited so that the most slowly growing group in the food chain, the obligate proton-reducing acetogens, can still excrete hydrogen at a maximum rate, but at the same time hydrogen accumulation >10–5 atm is prevented by active methanogenesis or sulphate reduction (Bryant, 1979).

Whereas hydrogen seemed to be the most sensitive regulator of anaerobic degradation, formate (Bleicher and Winter, 1994), acetate, or other fatty acids accumulated to much higher concentrations (McInerney, 1988) but did not repress anaerobic degradation. If hydrogen accumulated due to an oversupply or to inhibition of methanogens, anaerobic biodegradation was disturbed successively in different stages. Initially, in the acetogenic stage a-oxidation of fatty acids and alcohols failed, leading to an accumulation of these acid metabolites. Later, the spectrum of metabolites of the fermentative flora changed towards more reduced products like ethanol, lactate, propionate, and *n*butyrate, leading to an even higher concentration of volatile fatty acids and a further decrease in pH. At pH values below 6.5, methanogenic reactions were almost completely prevented. At this stage acidification with a rich spectrum of reduced products still proceeded.

Anaerobic Degradation of Carbohydrates in Wastewater

Carbohydrates are homo- or heteropolymers of hexoses, pentoses, or sugar derivatives, which occur in soluble form or as particles, forming grains or fibres of various sizes. In some plants, starch forms grains up to 1 mm in diameter, which is 1000 times the diameter of bacteria. Starch metabolism by bacteria requires hydrolytic cleavage by amylases to form soluble monomers or dimers, since only soluble substrates can be taken up and metabolised.

The anaerobic degradation of biopolymers in general and of cellulose in particular can be divided into hydrolytic, fermentative, acetogenic, and methanogenic phases. Hydrolysis and fermentation of the hydrolysis products can be catalysed by the same trophic group of microorganisms. The distinction of the two phases is of more theoretical than practical relevance. Concerning reaction rates in a methane fermenter that is fed with a particulate substrate, the rate-limiting step is hydrolysis rather than the subsequent fermentation of the monomers, if acetogenesis and methanogenesis proceed faster. The hydrolysis rates of polymers can be very different.

Hemicellulose and pectin are hydrolysed ten times faster than lignin-encrusted cellulose (Buchholz *et al.*, 1986, 1988). In the acidification reactor of a two-stage anaerobic process, hydrolysis of polymers to monomers is normally slower than fermentation of monomers to fatty acids and other fermentation products. For this reason, no sugar monomers can be detected during steady-state operation. In the methane reactor, a-oxidation of fatty acids, especially of propionate or *n*-butyrate, is the ratelimiting step (Buchholz *et al.*, 1986). Fatty acid degradation is the slowest reaction overall in a two-stage methane reactor fed with carbohydrate-containing wastewater from sugar production. Thus, the methane reactor has to be larger than the acidification reactor to permit longer hydraulic retention times.

The rate of cellulose degradation depends strongly on the state of the cellulose in the wastewater. If cellulose is lignin-encrusted, lignin prevents access of cellulases to the cellulose fibres. If cellulose is mainly in a crystalline form, cellulases can easily attach to it, and then hydrolysis can be a relatively fast process. At increasing loading in an anaerobic reactor fed with crystalline cellulose, acetogenesis became the rate-limiting process, leading to propionate and butyrate formation (Winter and Cooney, 1980). In decaying plant material, cellulose is very often lignin-encrusted. Due to the highly restricted access to these complexes by cellulases, hydrolysis of cellulose is the rate-limiting step in its degradation to methane and CO_2.

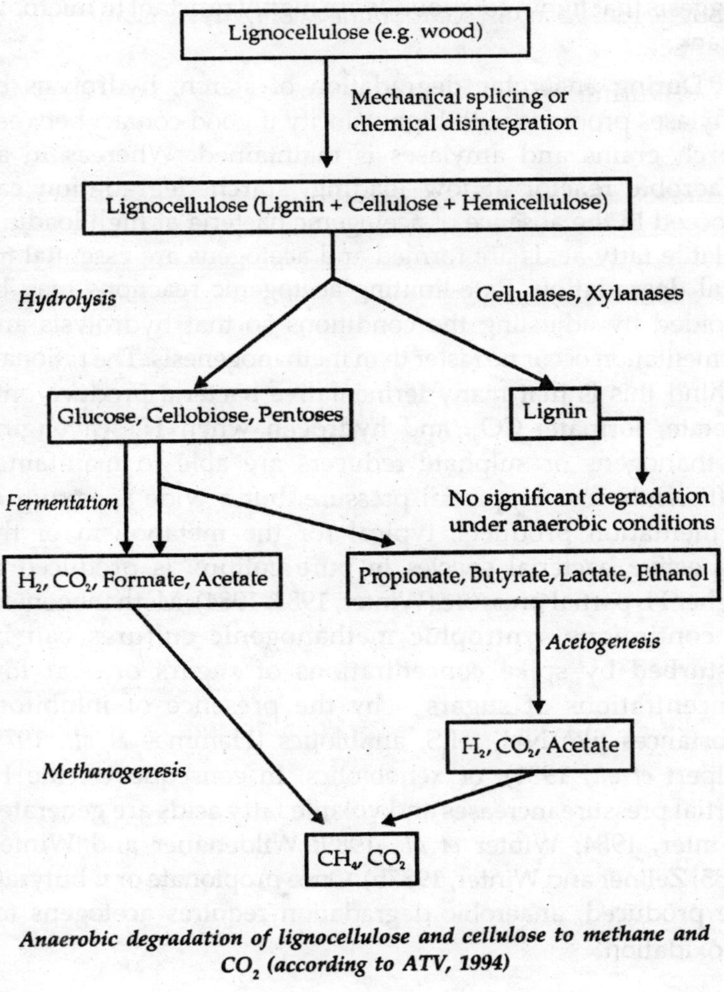

Anaerobic degradation of lignocellulose and cellulose to methane and CO_2 (according to ATV, 1994)

Whether microorganisms are capable of degrading lignin under anaerobic conditions is still under discussion. In a natural environment without time limitation, lignin was reported to be degraded anaerobically (Colberg and Young, 1985; Colberg, 1988). However, since these results were based on long-term experiments performed *in situ* and anaerobiosis was not controlled, it remains doubtful, whether the small amount of lignin that disappeared was really degraded under strictly anaerobic conditions. The occurrence of coal and fossil oil suggests that lignin compounds are highly resistant to microbial attack.

During anaerobic degradation of starch, hydrolysis by amylases proceeds with high velocity if good contact between starch grains and amylases is maintained. Whereas in an anaerobic reactor at low loading, starch degradation can proceed in the absence of acetogenic bacteria at high loading, volatile fatty acids are formed and acetogens are essential for total degradation. Rate-limiting acetogenic reactions may be avoided by adjusting the conditions so that hydrolysis and fermentation occur no faster than methanogenesis. The rationale behind this is that many fermentative bacteria produce only acetate, formate, CO_2, and hydrogen when H_2-scavenging methanogens or sulphate reducers are able to maintain a sufficiently low H_2 partial pressure, but a wide spectrum of fermentation products, typical for the metabolism of the respective bacterial species in pure culture, is produced at higher H_2 partial pressure (Winter, 1983, 1984). Methanogenesis in continuous syntrophic methanogenic cultures can be disturbed by spike concentrations of sugars or – at low concentrations of sugars – by the presence of inhibitory substances like NH_2, H_2S, antibiotics (Hammes *et al.*, 1979; Hilpert *et al.*, 1981), or xenobiotics. In consequence, the H_2 partial pressure increases and volatile fatty acids are generated (Winter, 1984; Winter *et al.*, 1989; Wildenauer and Winter, 1985; Zellner and Winter, 1987b). Once propionate or *n*-butyrate are produced, anaerobic degradation requires acetogens for β-oxidation.

Anaerobic Degradation of Protein

Proteins are biological macromolecules, either soluble or solid (e.g., feathers, hair, nails). Outside the cell at an acid pH or in the presence of enzymes, soluble proteins precipitate, e.g., precipitation of casein by addition of rennet enzyme. The reaction sequences necessary for protein degradation in a methanogenic ecosystem. Hydrolysis of precipitated or soluble protein is catalysed by several types of proteases that cleave membrane-permeable amino acids, dipeptides, or oligopeptides. In contrast to the hydrolysis of carbohydrates, which proceeds favourably at a slightly acid pH, optimal hydrolysis of proteins requires a neutral or weakly alkaline pH (McInerney, 1988). In contrast to the fermentation of carbohydrates, which lowers the pH due to volatile fatty acid formation, fermentation of amino acids in wastewater reactors does not lead to a significant pH change, due to acid and ammonia formation. Acidification of protein-containing wastewater proceeds optimally at pH values of 7 or higher (Winterberg and Sahm, 1992), and ammonium ions together with the CO_2–bicarbonate–carbonate buffer system stabilise the pH.

Acetogenesis of fatty acids from deamination of amino acids requires a low H_2 partial pressure for the same reasons as for carbohydrate degradation. This can be maintained by a syntrophic interaction of fermentative, protein-degrading bacteria and acetogenic and methanogenic or sulphate-reducing bacteria. Except for syntrophic interaction of amino acid-degrading bacteria with methanogens for maintenance of a low H_2 partial pressure, clostridia and presumably also some other sludge bacteria may couple oxidative and reductive amino acid conversion via the Stickland reaction.

One amino acid, e.g., alanine (Eq. 4) is oxidatively decarboxylated and the hydrogen or reducing equivalents produced during this reaction are used to reductively convert another amino acid, e.g., glycine, to acetate and ammonia (Eq. 5).

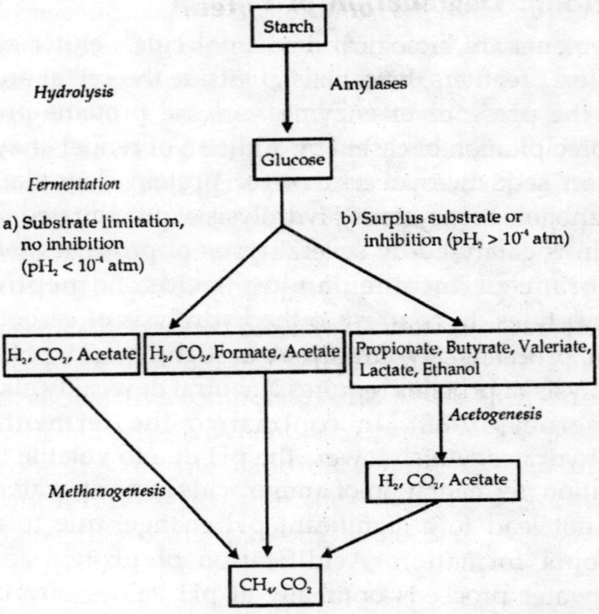

Anaerobic degradation of starch under low-and high-loading conditions

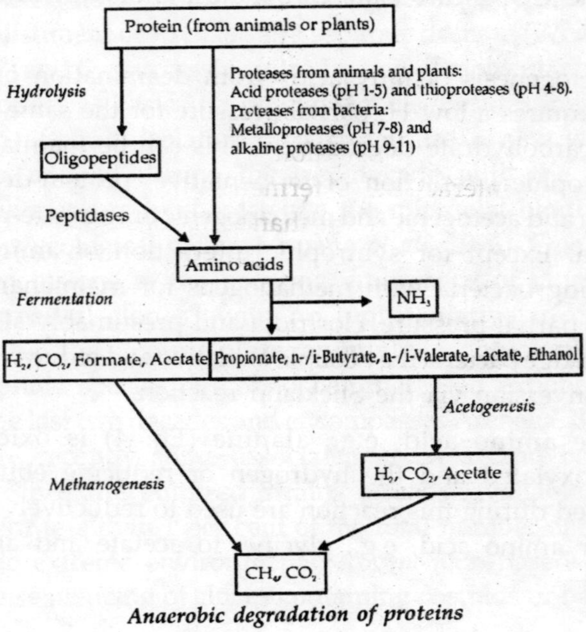

Anaerobic degradation of proteins

$$CH_3\text{–}CHNH_2\text{–}COOH + 2\ H_2O \rightarrow CH_3\text{–}COOH + CO_2 + NH_3 + 2\ H_2 \quad (4)$$

$\Delta\ G^{o\prime} = +7.5$ kJ mol^{-1}

$$2\ CH_2NH_2\text{–}COOH + 2\ H_2 \rightarrow 2\ CH_3COOH + 2\ NH_3 \quad (5)$$

$\Delta\ G^{o\prime} = -38.9$ kJ mol–1

For complete degradation of amino acids in an anaerobic system therefore, a syntrophism of amino acid-fermenting anaerobic bacteria with methanogens or sulphate reducers is required (Wildenauer and Winter, 1986; Winter *et al.*, 1987; Orlygsson *et al.*, 1995). If long-chain amino acids are deaminated (Eqs. 6–9), fatty acids such as propionate, *i*-butyrate, or *i*-valerate are formed directly. The fatty acids require acetogenic bacteria for their degradation.

$$\text{valine} + 2\ H_2O \rightarrow i\text{-butyrate} + CO_2 + NH_3 + 2\ H_2 \quad (6)$$

$$\text{leucine} + 2\ H_2O \rightarrow i\text{-valerate} + CO_2 + NH_3 + 2\ H_2 \quad (7)$$

$$i\text{-leucine} + 2\ H_2O \rightarrow \text{2-methylbutyrate} + CO_2 + NH_3 + 2\ H_2 \quad (8)$$

$$\text{glutamate} + 2\ H_2O \rightarrow \text{propionate} + 2\ CO_2 + NH_3 + 2\ H_2 \quad (9)$$

In contrast to carbohydrate degradation, where the necessity for propionate-and butyrate-degrading acetogenic bacteria can be circumvented by substrate limitation, during protein degradation these fatty acids are a product of deamination, and their formation cannot be avoided by maintaining a low H_2 partial pressure.

In the methanogenic phase there is no difference in methanogenic activity whether carbohydrates or proteins are fermented, except that the methanogens in a reactor fed with protein need to be more tolerant to ammonia and higher pH.

Anaerobic Degradation of Neutral Fats and Lipids

Fats and lipids are another group of biopolymers that contribute significantly to the COD in sewage sludge, cattle and swine manures, and wastewater from the food industry, e.g., slaughterhouses or potato chip factories (Winter *et al.*, 1992; Broughton *et al.*, 1998). To provide a maximum surface for hydrolytic cleavage by lipases or phospholipases, solid fats, lipids, or oils must be emulsified. Glycerol and saturated and unsaturated fatty acids (palmitic acid, linolic acid, linolenic acid, stearic acid, etc.) are formed from neutral fats. Lipolysis of phospholipids generates fatty acids, glycerol, alcohols (serine, ethanolamine, choline, inositol), and phosphate. Lipolysis of sphingolipids generates fatty acids and amino alcohols (e.g., sphingosine), and lipolysis of glycolipids generates fatty acids, amino alcohols, and hexoses (glucose, galactose).

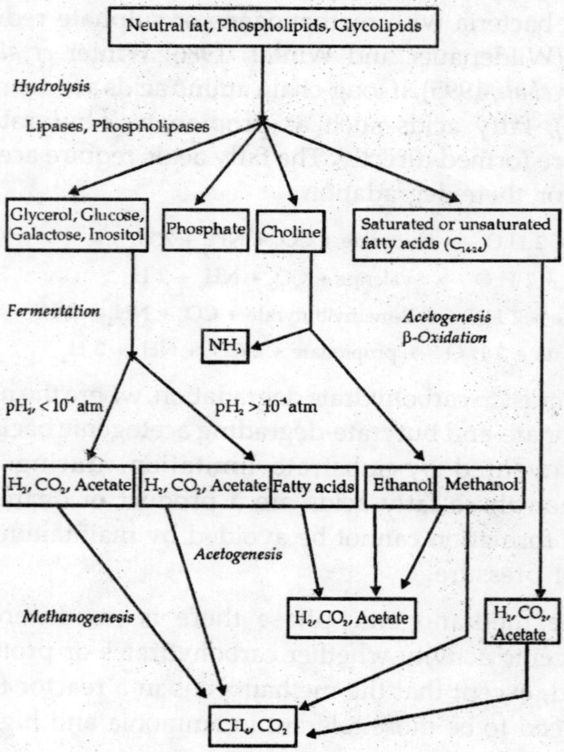

Anaerobic degradation of facts

Sugar moieties and glycerol can be degraded to methane and CO_2 by interaction between fermentative and methanogenic bacteria in low-loaded systems or by cooperation between fermentative, acetogenic, and methanogenic bacteria in high-loaded systems. The long-chain fatty acids are degraded by acetogenic bacteria by a-oxidation to acetate and molecular hydrogen. If acetate and molecular hydrogen accumulate, the anaerobic digestion process is inhibited (Hanaki *et al.*, 1981). Odd-numbered fatty acids are degraded to acetate, propionate, and hydrogen, and even-numbered fatty acids to acetate and hydrogen (Bryant, 1979). Only at a very low H_2 partial pressure, which can be maintained by hydrogen-utilising methanogens or sulphate reducers, is β-oxidation of at least *n*-butyrate or propionate exergonic. Methanol, ethanol, and ammonia are formed from choline. After hydrolysis, fermentation, and acetogenesis of the fat components in the methanogenic phase acetate, CO_2, and hydrogen are converted to biogas. All subsequent intracellular reactions can be influenced by syntrophic interaction via interspecies hydrogen transfer, except for the extracellular initial lipase reaction.

Carbohydrates, proteins, fats, and biogenic oils can also be degraded anaerobically under thermophilic conditions. The overall degradation scheme is the same, but the microorganism populations are different (e.g., Winter and Zellner, 1990). Thermophilic fat degradation is becoming more important in practice, since waste fat from fat separators and fat flotates of the food industry are often cofermented in agricultural biogas plants. Since, for hygienic reasons, the input material must be autoclaved, a thermophilic process should be used, keeping the fat in a melted, soluble form for more effective metabolism. Biogas plants with cofermentation of waste fat residues are considered waste treatment systems and must be designed to meet the hygienic demands relevant to treatment of the respective waste.

Competition of Sulphate Reducers with Methanogens in Methane Reactors

Municipal wastewater or wastewater from sugar production, slaughterhouses, breweries, etc., normally contains less than 200 mg L–1 of sulphate. If sulphuric acid is used, e.g., to clean stainless steel containers and pipes in the dairy industry or to maintain an acid pH in bioreactors for bakers' yeast or citric acid production, or if ammonium sulphate is used to inhibit metabolic routes in bakers' yeast for the production of biochemicals, the wastewater contains large amounts of sulphate. Sulphite-containing wastewater is also generated by the starch and cellulose industry during bleaching of the raw products.

If these wastewaters are subjected to anaerobic treatment, the methanogenic bacteria must compete with sulphate reducers for the hydrogen equivalents from COD degradation (Omil *et al.*, 1998). In anaerobic digesters sulphate reduction is the favoured reaction, due to the higher affinity of sulphate reducers for reducing equivalents or hydrogen, leading to sulphide production at the expense of reduced biogas formation.

Sulphide subsequently forms heavy metal precipitates and, if still present in larger amounts, remains in solution and may be toxic to acetogens and methanogens. Some hydrogen sulphide leaves the reactor with the biogas, which may contain up to a few per cent of H_2S. Then gas purification is necessary before the gas can be used as fuel for gas engines. Hydrogen sulphide in biogas causes not only odour problems but also corrosion of fermenters and pipes.

Methanogenesis of dissolved and particulate organic material in sulphate-rich wastewater is possible, if acidogenesis and sulphate reduction (stripping of H_2S together with CO_2) is separated from methanogenesis in a phased or a staged process (Lens *et al.*, 2002). The higher affinity of sulphate reducers than methanogens for reducing equivalents can also be used for sulphate and heavy metal ion removal from wastewater in the first stage of a two-stage anaerobic process (Elferink *et al.*, 1994).

In comparison to methanogens, which have a rather restricted substrate spectrum, sulphate reducers are metabolically more versatile. Sulphate reducers can utilise polymers such as starch, monomers such as sugars, fatty acids, formate, aliphatic and aromatic compounds, as well as molecular hydrogen (Widdel, 1988) to generate reducing equivalents for sulphate reduction (Eq. 10):

$$8\ (H) + 2\ H^+ + SO_4^{2-} \rightarrow H_2S + 4\ H_2O \qquad (10)$$

Undissociated hydrogen sulphide is toxic to both methanogens and sulphate reducers. A 50 per cent inhibition of methanogenesis was observed at a total sulphide concentration of 270 mg L^{-1} (Oleszkiewicz *et al.*, 1989), whereas 85 mg L^{-1} sulphide inhibited sulphate reducers (McCartney and Oleszkiewicz, 1991). For stable methanogenesis, no more than 150 mg L^{-1} sulphide should be accumulated (Speece *et al.*, 1986). H_2S toxicity can be avoided by intensive flushing with biogas to strip the H_2S for adsorption onto $Fe_2(OH)_3$. If large amounts of H_2S are formed during anaerobic treatment, essential heavy metal ions for methanogens (Fe, Ni, Mo, Co, etc.) may be precipitated as sulphides, which may lead to deficiencies in heavy metal bioavailability for the wastewater population.

The metabolites of the fermentative and acetogenic phase of anaerobic wastewater treatment systems, mainly acetate and $CO_2 + H_2$, are substrates for methanogens and for sulphate reducers. If wastewater with a high sulphate concentration is treated in a methane reactor, the population may gradually shift from hydrogenotrophic methanogens towards hydrogenotrophic sulphate reducers, due to a more favourable *Ks* value for hydrogen of the sulphate reducers. For pure cultures of *Desulphovibrio* sp. the *Ks* value for hydrogen is 1 μM, but for hydrogenotrophic methanogens it is only 6 μM (Kristjansson *et al.*, 1982). For acetate, such a dramatic disadvantage of methanogens was not observed. The affinities of sulphate reducers and of methanogenic bacteria (valid at least for *Methanosaeta* sp.) for hydrogen are in the same range, 0.2–0.4 μM (Elferink *et al.*, 1994).

Amount and Composition of Biogas During Fermentation of Carbohydrates, Proteins, and Fats

From an economic point of view, the specific biogas

amounts and the biogas quality from anaerobic treatment of wastewater and sludge are important process parameters.

Therefore, a basis for prediction of amount and composition of biogas was elaborated as early as the 1930s (Buswell and Neave, 1930; Buswell and Sollo 1948; Buswell and Mueller, 1952). If the elemental composition of a wastewater is known, the theoretical amount and composition of the biogas can be predicted with the Buswell equation (Eq. 11). The actual biogas amount is lower and can be calculated by including a correction factor for the degree of degradability, the pH (which influences CO2 absorption), and a 5 per cent–10 per cent discount for biomass formation.

$$C_cH_hO_oN_nS_s + 1/4\ (4c-h-2o+3n+2s)\ H_2O \rightarrow 1/8\ (4c-h+2o+3n+2s)\ CO_2 + 1/8\ (4c+h-2o-3n-2s)\ CH_4 + nNH_3 + sH_2S \quad (11)$$

According to the Buswell equation (Eq. 11), for the anaerobic treatment of a wastewater with carbohydrates as pollutants, the gas composition should theoretically be 50 per cent methane and 50 per cent CO_2 (Eq. 12):

$$1\ \text{mol}\ C_6H_{12}O_6 \rightarrow 3\ \text{mol}\ CH_4 + 3\ \text{mol}\ CO_2 \quad (12)$$

Since CO_2 is increasingly soluble in water with decreasing temperature and increasing pH, CO_2 reacts to form bicarbonate/ carbonate, and the biogas may contain more than 80 per cent methane. The total amount of gas is then diminished by the amount of CO_2 that is absorbed and solubilised in the liquid. From a fat-and protein-containing wastewater, theoretically more than 50 per cent methane can be generated.

Nitrogen Removal During Wastewater Treatment

Nitrogen-containing substances in wastewater are inorganic or organic. Together with phosphates, they represent the main source for eutrophication of surface water.

For this reason they must be eliminated together with the organic carbon during wastewater treatment. Whereas phosphates form insoluble precipitates with many heavy metal ions and the precipitates can be separated by sedimentation or flotation, all nitrogen compounds, except for magnesium ammonium phosphate, are easily soluble in water and thus cannot be removed chemically by precipitation. For biological removal of amino nitrogen and of heterocyclic nitrogen compounds, their conversion to ammonia in an aerobic or anaerobic treatment process is the first step. Then ammonia must be nitrified, and the nitrate denitrified to yield nitrogen.

Thus, depending on the kind of nitrogen compounds present in wastewater, nitrogen removal requires up to three processes in sequence: ammonification, nitrification, and denitrification.

The major portion of nitrogen compounds in municipal wastewater are reduced nitrogen compounds such as ammonia, urea, amines, amino acids, and proteins.

Oxidised nitrogen compounds such as nitrate and nitrite normally are not present at all or not in relevant amounts. Nitrate and nitrite may, however, represent the main nitrogen load in wastewater of certain food or metal industry branches (Gensicke *et al.*, 1998; Zayed and Winter, 1998).

Amount of biogas, biogas composition, and energy content

Substrate	*Amount* ($cm^3\ g^{-1}$)	*Composition* $\%CH_4$	$\%\ CO_2$	*Energy content* ($kWh \times m^{-3}$)[d]
Carbohydrates[a]	746.7	50	50	4.95
Fats[b]	1434	71	29	7.02
Proteins[c]	636	60	40	5.93

a Calculated for hexoses.

b Calculated for triglycerides containing glycerol plus 3 mol palmitic acid.

c Calculated for polyalanine and reaction of ammonia to $(NH_4)2CO_3$.

d at standard conditions

The ammonia in raw municipal wastewater is mainly derived from urine and is formed in the sewer system by enzymatic cleavage of urea by ureases (Eq. 13):

$$NH_2CONH_2 + H_2O \rightarrow CO_2 + 2NH_3 \qquad (13)$$

The residence time of the wastewater in the sewer system normally is not long enough for a significant contribution of ammonia from other sources, e.g., proteolysis and deamination of amino acids.

Ammonification

The main organic nitrogen compounds in municipal wastewater are heterocyclic compounds (e.g., nucleic acids) and proteins. Proteolysis and degradation of amino acids leads to liberation of ammonia by the various mechanisms of ammonification (Rheinheimer *et al.*, 1988), including hydrolytic, oxidative, reductive, and desaturative deamination (Eqs. 14-17, respectively)

$$R\text{–}NH_2 + H_2O \rightarrow R\text{–}OH + NH_3 \quad (14)$$

$$R\text{–}CHNH_2COOH + H_2O \rightarrow R\text{–}CO\text{–}COOH + 2\ (H) + NH_3 \quad (15)$$

$$R\text{–}CHNH_2\text{–}COOH + 2\ (H) \rightarrow R\text{–}CH_2\text{–}COOH + NH_3 \quad (16)$$

$$R\text{–}CH_2\text{–}CHNH_2\text{–}COOH \rightarrow R\text{–}CH{=}CH\text{–}COOH + NH_3 \quad (17)$$

A significant amount of ammonia from urea cleavage or from ammonification of amino acids is assimilated in aerobic treatment processes for growth of bacteria (surplus sludge formation). It can be estimated that bacteria consist of roughly 50 per cent protein and that the nitrogen content of protein is about 16 per cent. Thus, for synthesis of 1 g of bacterial biomass, about 0.08 g of ammonia-N is required. To eliminate ammonia that is not used for cell growth during wastewater treatment, it must first be nitrified and then denitrified to molecular nitrogen or anaerobically oxidised with nitrite.

Nitrification of Ammonia

Autotrophic Nitrification: Autotrophic nitrifiers are aerobic microorganisms oxidising ammonia via nitrite (Eq. 18) to nitrate (Eq. 19). Organisms catalysing nitrification (Eq. 18) belong to the genera *Nitrosomonas*, *Nitrosococcus*, *Nitrosolobus*, *Nitrosospira*, and *Nitrosovibrio*, organisms catalysing nitration (Eq. 19) include, e.g., members of the genera *Nitrobacter, Nitrococcus,* and *Nitrospira.*

$$NH_4^+ + 1.5\ O_2 \rightarrow NO_2^- + 2\ H^+ + H_2O \quad (18)$$

$$NO_2^- + 0.5\ O_2 \rightarrow NO_3^- \quad (19)$$

Ammonia oxidation to nitrite or nitrite oxidation to nitrate are energy-yielding processes for the autotrophic growth of the nitrifying bacteria. CO_2 is assimilated via the Calvin cycle. Since the positive redox potential of the oxidisable nitrogen compounds is not low enough to form $NADH_2$ for CO_2 reduction, $NADH_2$ must be formed by an energy-consuming reverse electron transport. For this reason, the growth yield of nitrifying bacteria is low. *Nitrosomonas* sp., e.g., must oxidise 30 g NH_3 to form 1 g of cell dry mass (Schlegel, 1992).

The oxidation of ammonia by nitrifiers is initiated by an energy-neutral monooxygenase reaction that yields hydroxylamine (Eq. 20):

$$NH_3 + XH_2 + O2 \rightarrow NH_2OH + X + H_2O \quad (20)$$

Then hydroxylamine is further oxidised, presumably via nitroxyl (Eq. 21) to nitrite (Eq. 22), which is the energy-yielding reaction during nitrification. For microbial oxidation of 1 mg NH^+_4 - nitrogen to nitrite, 3.42 mg O_2 is required, and for the oxidation of 1 mg NO^-_2 -nitrogen to nitrate (Eq. 19), 1.14 mg O_2.

$$NH_2OH + X \rightarrow XH_2 + (NOH) \quad (21)$$

$$(NOH) + 0.5\ O_2 \rightarrow HNO_2 + \text{energy} \quad (22)$$

Due to their slow growth autotrophic nitrifiers cannot successfully compete with heterotrophic bacteria for oxygen. In a highly loaded activated sludge system, the autotrophic nitrifiers are overgrown by the heterotrophic sludge flora, which consume the oxygen. Ammonia oxidation starts only if the BOD5 concentration in the wastewater is <110 mg L^{-1} (Wild *et al.*, 1971). During nitrification of ammonia the alkalinity of the wastewater increases slightly, due to CO_2 consumption for autotrophic growth (pH increase), but in a counter reaction it drastically decreases due to nitric acid formation from ammonia (pH decrease from above neutral to acidic). If the buffer capacity of the wastewater is weak, the pH drops far below 7 and thus prevents further nitrification by autotrophic nitrifiers (Rheinheimer *et al.*, 1988).

Heterotrophic Nitrification

Some bacteria of the genera *Arthrobacter, Flavobacterium,* and *Thiosphaera* are able to catalyse heterotrophic nitrification of nitrogen-containing organic substances (Eq. 23):

$$R\text{–}NH_2 \rightarrow R\text{–}NHOH \rightarrow R\text{–}NO \rightarrow NO_3^- \quad (23)$$

Heterotrophic nitrifiers oxidise reduced nitrogen compounds, such as hydroxylamine and aliphatic and aromatic nitrogen-containing compounds, but in contrast to autotrophic nitrification, no energy is gained by nitrate formation. For this reason an organic substrate must be respired to satisfy the energy metabolism (Schlegel, 1992).

Some heterotrophic nitrifiers can denitrify nitrate or nitrite under aerobic growth conditions. The nitrogen metabolism of *Thiosphaera pantotropha* and *Paracoccus denitrificans* are well documented examples (Stouthamer *et al.*, 1997). These organisms express a membrane-bound nitrate reductase under anoxic growth conditions, which works only in the absence of molecular oxygen. Under aerobic growth conditions a periplasmic nitrate reductase is expressed, which catalyses nitrate reduction at least to the state of nitrous oxide.

Denitrification: Nitrate Removal from Wastewater

Many aerobic bacteria seem to be able to switch their oxidative metabolism to nitrate respiration. Similar to oxygen respiration, the nitrate respiration of heterotrophic bacteria requires a complex carbon source as an electron source for denitrification (e.g., Eqs. 24 and 25). Denitrification (Eq. 26) starts with the reduction of nitrate to nitrite by membrane-bound nitrate reductase A (a). Then a membrane-bound nitrite reductase (b) catalyses NO formation. Finally, NO reductase

(c) and N_2O reductase (d) form N_2. The theoretical stoichiometry of denitrification with methanol or acetate as a carbon source is shown in Eqs. 27 and 28. For practical application a surplus of carbon source must be supplied, since the wastewater is not free of oxygen, and part of the carbon source is respired until anoxic conditions are achieved.

To evaluate the stoichiometry of nitrate to organic compounds for denitrification with a complex carbon source, the oxidation–reduction state of the carbon substrates and the oxygen concentration in the wastewater should be known. In wastewater treatment plants more than 2.85 g COD is required for reduction of 1 g NO_3^- –N (Bernet *et al.*, 1996).

$$CH_3OH + H_2O \longrightarrow CO_2 + 6\ H^+ + 6\ e^- \quad (24)$$

$$CH_3COOH + 2\ H_2O \longrightarrow 2\ CO_2 + 8\ H^+ + 8\ e^- \quad (25)$$

$$NO_3^- \xrightarrow[a]{2e^-} NO_2 \xrightarrow[b]{e^-} NO \xrightarrow[c]{e^-} 0.5N_2O \xrightarrow[d]{e^-} 0.5N_2 \quad (26)$$

$$5\ CH_3OH + 6\ HNO_3 \longrightarrow 5\ CO_2 + 3\ N_2 + 13\ H_2O \quad (27)$$

$$5\ CH_3\text{–}COOH + 8\ HNO_3 \longrightarrow 10\ CO_2 + 4\ N_2 + 14\ H_2O \quad (28)$$

Instead of nitrate, many denitrifying bacteria can use NO_2^-, NO, or N_2O as terminal electron acceptors. Alternatively, they may release these intermediates during denitrification of nitrate under unfavourable conditions, as was observed in soil (Conrad, 1996). If, e.g., surplus nitrate is supplied and hydrogen donors are not sufficiently available, NO and N_2O can be formed (Schon *et al.*, 1994). Another condition for N_2O formation is a pH below 7.3, at which nitrogen oxidoreductase is inhibited (Knowles, 1982).

Except for dissimilatory nitrate reduction, many aerobic and anaerobic bacteria are capable of assimilatory nitrate reduction to supply the cells with ammonia for growth (Eqs. 29 and 30). However, the enzymes for nitrate assimilation are expressed only at concentrations of ammonia <1 mM.

$$NO_3^- \xrightarrow[a]{2e^-} NO_2^- \xrightarrow[b]{2e^-} HNO \xrightarrow[c]{2e^-} NH_2OH \xrightarrow[d]{2e^-} NH_3 \quad (29)$$

$$HNO_3 + 8(H) \longrightarrow NH_4^+ OH^+ + 2H_2O \tag{30}$$

Here, nitrate reductase B (enzyme *a* in Eq. 29) reduces nitrate to nitrite, which is then reduced to ammonia by the nitrite reductase complex (enzyme b in Eq. 29).

Whereas nitrate reduction by the oxygen-sensitive, membrane-bound enzyme nitrate reductase. A conserves energy, no energy conservation is possible in the reaction catalysed by the soluble enzyme nitrate reductase B, which is not repressed by oxygen. For details on the cell biology and the molecular basis of denitrification, please refer to Zumft (1997).

Combined Nitrification and Denitrification

A strict separation of the reactions participating in aerobic, autotrophic nitrification and in anoxic, heterotrophic denitrification is not required, as concluded from N15-tracer experiments (Kuenen and Robertson, 1994). Autotrophic ammonia oxidisers seem to be able to produce NO, N_2O, or N_2 from nitrite if oxygen is limited, and ammonia as well as nitrite oxidisers can be isolated from anaerobic reactors (Kuenen and Robertson, 1994). *Nitrosomonas europaea* can use nitrite as an electron acceptor and pyruvate as an energy source under anoxic, denitrifying growth conditions (Abeliovich and Vonshak, 1992). In addition, several strains of *Nitrobacter* sp. were reported to denitrify during anoxic, heterotrophic growth (Bock *et al.*, 1986).

Aerobic denitrification by *Thiosphaera pantotropha* was first described by Robertson and Kuenen (1984). *T. pantotropha* respires molecular oxygen and denitrifies nitrate simultaneously, provided that suitable electron acceptors are available. The conversion rate of acetate as electron donor with nitrate as electron acceptor was twice as high when the concentration of molecular oxygen was <30 per cent of air saturation compared to 30–80 per cent air saturation (Robertson *et al.*, 1988). Respiration, simultaneous nitrification, and denitrification were observed in the presence of oxygen and ammonia. During heterotrophic denitrification by *T. pantotropha*, ammonia is first oxidised to hydroxylamine by an ammonia monooxygenase, with ubiquinone serving as electron donor. Hydroxylamine is subsequently oxidised to nitrite by a hydroxylamine oxidoreductase. During coupled nitrification–denitrification, 3 of the 4 electrons from the oxidation of hydroxylamine are used for reduction of nitrite to nitrogen and are not available to the electron transport reaction catalysed by cytochrome oxidase. Regeneration of ubiquinone is mediated by electrons that are generated during oxidation of an organic substrate (heterotrophic nitrification).

Conversion rates of ammonia by heterotrophic nitrifiers such as *T. pantotropha* (Kuenen and Robertson, 1994) are smaller than those of autotrophic nitrifiers (35.4 for *T. pantotropha* versus 130–1550n mol NH_3 min^{-1} mg^{-1} dry weight for *Nitrosomonas* sp.). If, however, the higher population density of heterotrophic nitrifiers, resulting from higher growth rates, is taken into consideration, the specific conversion rates are in a similar range. Heterotrophic nitrifiers can, in addition to their nitrifying capability, denitrify nitrite or nitrate to molecular nitrogen. In many wastewater treatment plants autotrophic nitrifiers may exist, producing nitrite from ammonia under moderately aerobic conditions; the ammonia is then converted to nitrate and/or reduced to nitrogen by heterotrophic nitrifiers in the presence of a suitable carbon source. In practice, the only disadvantage of heterotrophic nitrification is more surplus sludge generation for final disposal.

Anaerobic Ammonia Oxidation (Anammox)

To date, two anaerobic ammonia-oxidising bacteria species have been isolated and preliminarily classified in the order Planctomycetales as candidatus *Brocardia anammoxidans* and candidatus *Kuenenia stuttgartiensis*. Under anaerobic conditions ammonia is oxidised to nitrite in membrane-bound intracytoplasmic anammoxosoms (Schmidt *et al.*, 2003).

In the Anammox process ammonia is oxidised to nitrogen, with nitrite serving as electron acceptor (Van de Graaf *et al.*, 1995). Use of N^{15} isotopes indicated that nitrite is reduced to hydroxylamine (Eq. 31), which then reacts with ammonia to yield hydrazine (N_2H_4, Eq. 32). By oxidation of hydrazine to molecular nitrogen (Eq. 33), four reducing equivalents are generated, which are required for nitrite reduction to hydroxylamine (Van de Graaf *et al.*, 1997).

$$2\ HNO_2 + 4\ XH_2 \longrightarrow 2\ NH_2OH + 2\ H_2O + 4X \qquad (31)$$

$$2\ NH_2OH + 2\ NH_3 \longrightarrow 2\ N_2H_4 + 2\ H_2O \qquad (32)$$

$$2\ N_2H_4 + 4\ X \longrightarrow 2\ N_2 + 4\ XH_2 \qquad (33)$$

Since the redox state is balanced in the above reactions, reducing equivalents for CO_2 reduction by the autotrophic microorganisms must be generated by oxidation of nitrite to nitrate (Eq. 34):

$$HNO_2 + H_2O + NAD \longrightarrow HNO_3 + NADH_2 \qquad (34)$$

Per mol of ammonia, 0.2 mol nitrate is generated and 20 mg biomass is produced (Van de Graaf *et al.*, 1996). The Anammox process seems to be suitable for nitrogen removal in ammonia-rich effluents of anaerobic reactors that are fed with wastewater rich in TKN compounds. Nitrogenous compounds can be eliminated from this wastewater by a combination of nitrification for nitrite supply and anaerobic ammonia oxidation (Strous *et al.*, 1997). The process is sensitive to high, toxic nitrite concentrations of 70–180 mg NL^{-1} (depending on the kind of biomass). Furthermore, due to the low growth rate of the anammox bacteria, a startup time of 100–150 days with activated sludge as inoculum seems to be necessary (Schmidt *et al.*, 2003).

In oxygen-limited environments (e.g. oxic–anoxic interfaces of biofilms or sludge flocs) aerobic and anaerobic ammonia oxidisers are natural partners. The ammonia is oxidised to nitrite and concomitantly the oxygen level is decreased. The anammox bacteria disproportionate the nitrite and the remaining ammonia to N_2. When ammonia is limited, the affinities of aerobic and anaerobic ammonia oxidisers may lead the natural partners to change to competitors (Schmidt *et al.*, 2002). Other examples of 'uncontrolled' anammox activities were observed in different wastewater treatment plants with uncharacterised high ammonium losses.

N-removal Processes

The metabolic versatility of nitrogen-converting bacteria offers new concepts for nitrogen removal processes to treat wastewater. A *partial* nitrification of ammonia to nitrite (nitritation) would have the advantage of saving a significant amount of oxygen for the nitrification process, as compared to nitratation of ammonia. However, nitrite is toxic and must be kept below toxic concentrations. In a subsequent denitrification process, fewer reducing equivalents from the degradation of different carbon sources, which often are a limiting factor in wastewater treatment plants, are then required. In the SHARON process, which is a single reactor system for rapid ammonia oxidation to nitrite and subsequent denitrification of the nitrite with methanol, nitrite oxidisers that would generate nitrate are outcompeted by the slowly metabilising ammonia oxidisers after a temperature shift towards higher temperatures, due to the higher temperature tolerance of the ammonia oxidisers, and by no sludge retention or sludge return in the reactor (Hellinga *et al.*, 1998).

Aeration was switched off periodically and denitrification of nitrite was initiated by adding methanol as a carbon source. The SHARON process in full scale was used to treat sludge liquor at the Rotterdam wastewater treatment plant (Mulder *et al.*, 2001). The partial nitrification of ammonia to nitrite led to an ammonia/nitrite ratio that was suitable for a subsequent Anammox process in which nitrite was denitrified to N_2 with ammonia instead of methanol as the electron donor. This process was finally established in full-scale at the Rotterdam wastewater treatment plant (Schmidt, 2003).

Partial ammonia oxidation to nitrite and anaerobic oxidation of ammonia in one aerated reactor is the concept behind the CANON process (completely autotrophic nitrogen removal over nitrite) (Van Lossdrecht and Jetten, 1997). Aerobic and anaerobic ammonia-oxidising bacteria cooperate as long as the nitrifying bacteria consume oxygen and create anoxic conditions for the anaerobic ammonia-oxidising bacteria.

Another possibility for removing ammonia in one single step without using an organic carbon source is the OLAND process (oxygen-limited nitrification and denitrification)

(Kuai and Verstraete, 1998). Conversion of ammonia to N_2 is catalysed by aerobic nitrifiers and anaerobic granular sludge as a source of *Planctomycetes* including anaerobic ammonia-oxidising bacteria (Pynaert *et al.*, 2004).

Other nitrogen-removal processes also exist, such as the NOx process (simultaneous nitrification and denitrification of *Nitrosomonas*-like microorganisms under fully oxic conditions) and aerobic deammonification (conversion of ammonia to N_2 and nitrate), which allow ammonia removal without COD. An overview of these new concepts was presented by Schmidt *et al.*, 2003.

Enhanced Biological Phosphate Removal

Based on an early observation that microorganisms take up more phosphate than required for cell growth, it was found that single cells accumulate polyphosphate in granules containing a few to several thousand phosphate units (Egli and Zehnder, 1994). Phosphate constitutes up to 12 per cent of the cell weight of polyphosphate-accumulating bacteria, whereas bacteria that do not accumulate polyphosphate contain only 1 per cent–3 per cent phosphate (van Loosdrecht *et al.*, 1997b). The accumulated polyphosphate is assumed to be an energy source for substrate assimilation during anaerobic growth conditions and poly-β-hydroxyalkanoate (PHA) synthesis. Evidence suggests that degradation of polyphosphate may be used to regulate intracellular pH under alkaline conditions (Seviour *et al.*, 2003). For biological phosphate removal by 'luxury phosphate uptake', the polyphosphate-accumulating bacteria in wastewater have to be subjected in sequence to an anaerobic and an aerobic environment. All previous attempts to isolate polyphosphate-accumulating bacteria (culture-dependent approach) led to recovery of *Acinetobacter* sp. which belongs to the γ -Proteobacteria and is the model organism for explanation of the polyphosphate accumulation mechanism found in textbooks. However, pure cultures of the polyphosphate-accumulating *Acinetobacter* sp. and wastewater communities seem to behave differently (van Loosdrecht *et al.*, 1997a). A major discrepancy is that *Acinetobacter* sp. do not play a dominant role in activated sludge systems, as deduced from a survey using specific gene probes (Wagner *et al.*, 1994). The culture-independent approach led to a mixed population for enhanced biological phosphorous removal (EBPR). The phosphorous-removing members (>10 per cent of the total population) were assigned to the β -Proteobacteria (Rhodocylus group, e.g. candidatus *Accumulibacter phosphatis*), -Proteobacteria and Actinobacteria (*Microlunatus phosphoruvorus*) and to members of the Cytophaga–Flexibacter–Bacteroides division (Seviour *et al.*, 2003).

In *Acinetobacter* sp. polyphosphate accumulation is closely interconnected with poly-β -hydroxybutyrate (PHB) and glycogen metabolism (van Loosdrecht *et al.*, 1997b). Under anaerobic conditions in the absence of nitrate, acetate is taken up by *Acinetobacter* cells, metabolised to β -hydroxybutyrate, polymerised to PHB, and stored intracellularly in inclusion bodies. Reducing equivalents for PHB formation from acetate are made available by glycolysis of glucose units from stored glycogen or are alternatively formed by other microorganisms for EBPR in an anaerobically operating TCA cycle. The energy for polymerisation comes from hydrolysis of the anhydride bonds of ATP and polyphosphate. To supply both the maintenance metabolism and the storage metabolism of *Acinetobacter* with sufficient energy under anaerobic conditions, anhydride bonds in polyphosphate are hydrolysed, and inorganic phosphate is excreted (phosphate resolubilisation) (Smolders *et al.*, 1994).Stored glycogen in microorganisms for EBPR may also provide ATP for PHA synthesis.

For storage of 1 mg fatty acids, about 0.6 mg orthophosphate is released (Danesh and Oleszkiewicz, 1997). Under aerobic conditions two situations may be prevalent. In wastewater that does not contain a suitable carbon source for respiration, *Acinetobacter* sp. hydrolyse PHB, which is stored in the granules, and respire the β -hydroxybutyrate to gain energy for growth, maintenance, formation of glycogen, and polymerisation of phosphate, which is taken up from the wastewater. In the presence of oxygen much more P_i is taken up by *Acinetobacter* sp. from the activated sludge flora than is released under anaerobic conditions ('luxury uptake').

Danesh and Oleszkiewicz (1997) reported that approximately 6 to 9 mg volatile fatty acids are required for biological removal of 1 mg phosphorous.

In wastewater that contains suitable carbon sources for *Acinetobacter* sp., these carbon sources are partially oxidised by the TCA cycle and are partially used for PHB formation (Kuba *et al.*, 1994). Since reducing equivalents are needed for PHB formation, less ATP is formed by oxidative phosphorylation. The lower ATP yield is compensated by the energy from hydrolysis of polyphosphate to P_i, which the cells use to form ADP from AMP. Two moles of ADP are disproportionated to AMP and ATP by adenylate kinase, and little Pi is released into the medium.

Except for obligately aerobic bacteria of the genus *Acinetobacter*, nitrate-reducing bacteria compete with microorganisms for EBPR for organic substrates, e.g., volatile fatty acids such as acetate. Nitrate-reducing bacteria also can eliminate phosphate from the aqueous environment (Kerrn-Jespersen and Henze, 1993). The simultaneous presence of obligately aerobic bacteria and nitrate-reducing bacteria in activated sludge may explain why, in the presence of nitrate in the anaerobic phosphate resolubilisation phase (oxygen absent), Pi does not accumulate in the medium.

The denitrifying, polyphosphate-accumulating bacteria apparently do not inhibit P_i excretion by *Acinetobacter*, but they take up phosphate and accumulate it as polyphosphate. This may explain why resolubilisation of phosphate by degradation of stored polyphosphate is not always observed in wastewater treatment plants for biological phosphate removal. Another possibility may be the inhibitory effect of nitric oxide on adenylate kinase, which is involved in polyphosphate degradation in *Acinetobacter* sp.

For biological phosphate accumulation, the ability of cells to store reserve material plays an essential role. Polymerisation of substrates and storage in intracellular granules may offer the advantage that, at times of substrate shortage or in an environment with low concentrations of substrates (at low k_m values), such organisms can survive much better than other heterotrophic bacteria that do not have effective strategies for substrate storage.

Biological phosphate elimination is a highly effective process leading to final concentrations in the wastewater of <0.1 mg L^{-1} phosphate. For optimal biological phosphate removal, the COD/P ratio in the wastewater should be about 20 g COD g^{-1} phosphate to permit good growth of the polyphosphate-accumulating bacteria. If biological phosphate elimination is used in combination with chemical precipitation, the minimum COD/P ratio should be 2 g COD g^{-1} phosphate (Smolders *et al.*, 1996) in processes with or without separation of primary sludge. If most of the particulate COD is separated in a primary sedimentation pond, ferrous ions should be added to supplement the biological phosphate removal with chemical precipitation.

If no presedimentation of sludge is available or the sedimentation efficiency is low, no ferrous ions have to be added (van Loosdrecht *et al.*, 1997b). A detailed description of the state of the art concerning biological phosphate elimination was given by Schon (1999).

Problems with the stability of the EBPR process in wastewater treatment plants were periodically reported. The reason for this may be the occurrence of microorganisms known as glycogen non-polyphosphate-accumulating organisms (GAOs).

These bacteria are selected under EBPR operation conditions (alternating anaerobic–aerobic phases during treatment) and compete with microorganisms for EBPR for the available volatile fatty acids (Blackall *et al.*, 2002). Filipe *et al.* (2001) suggested an important role of pH during anaerobic and aerobic operation of reactors: growth of GAOs is favoured at low pH, but at a pH of 7–7.5 the microorganisms for EBPR outcompete the GAOs. Further research is required to elucidate conditions under which GAOs are inhibited and growth of EBPR biomass is promoted.

Biological Removal, Biotransformation, and Biosorption of Metal Ions from Contaminated Wastewater

Whereas solid organic and inorganic material in wastewater or sludge can be removed by sedimentation, soluble organic pollutants and xenobiotics should be eliminated from the aqueous environment by microbial mineralisation or anaerobic degradation to gaseous products, with a varying portion (5 per cent – 50 per cent) being used as substrates for bacterial growth. Most of the inorganic components present in wastewater are soluble and are ionised. Trace amounts of many cations (e.g., Na^+, K^+, Ca^{2+}, Fe^{2+}, Ni^{2+}, Co^{2+}, Mn^{2+}, Zn^{2+}, etc.) and anions (e.g., PO_4^{3-}, Cl^-, S^{2-}) are essential micronutrients for bacterial growth. Other cations such as ammonia may also be required for bacterial growth, but the surplus amount must be oxidised to nitrate, and the nitrate denitrified to gaseous nitrogen, for N elimination from wastewater.

Under anaerobic conditions sulphate is reduced to sulphide, low amounts of which are required for growth of bacteria. The sulphide not required for growth is toxic for bacteria if present as H_2S in high concentrations. In anaerobic reactors at a slightly alkaline pH, most of the sulphide is precipitated as heavy metal sulphides; then it is harmless to microorganisms and their environment. To support detoxification, heavy metal ions may also be precipitated chemically.

Metal ion contaminants in wastewater can be removed by microorganisms by either a direct or indirect influence on the redox state of the metal ions or through biosorption of metal ions on the cell surface (Lovley and Coates, 1997). Some microorganisms have also developed resistance mechanisms against toxic metals by changing the oxidation state without supporting anaerobic growth. Certain bacteria, yeasts, fungi, and algae can actively accumulate intracellular metal ions against a gradient. The process of bioaccumulation of metal ions depends on living, metabolically active cells, whereas biosorption is a passive, energy-independent process that can be mediated also by inactive cell material. Biosorption of metal ions includes mechanisms such as ion exchange, chelation, matrix entrapment, and surface sorption (Unz and Shuttleworth, 1996).

After biosorption or active removal of metal ions from wastewater or contaminated soil, the heavy metal-containing biomass must be separated and incinerated or regenerated by desorption or remobilisation of the metals. As an example of successful biosorption, wastewater from the galvanising industry that contained 29 mg L^{-1} Zn and 10.5 mg L^{-1} Fe was purified in a three stage stirred-reactor cascade. With 15 g biomass, 7.5 L of wastewater was decontaminated (Brauckmann, 1997). The removal of metal ions by biologically catalysed changes in the redox state is an alternative. The altered speciation (valence status) of metals can lead to precipitation, solubilisation, or volatilisation of the metal ions.

Many bacteria use metal ions as electron acceptors for anaerobic respiration. Examples are the reduction of Fe^{3+}, Cr^{6+}, Mn^{4+}, Se^{6+}, As^{5+}, Hg^{2+}, Pb^{2+}, or U^{6+}. The dissimilatory metal-reducing bacteria can use H_2 or organic pollutants (e.g., xenobiotics) as electron donors and are capable of simultaneously removing organic and inorganic contaminants. Metal ions are reduced and precipitated by sulphide that is generated by sulphate-reducing bacteria.

Solubilisation of most heavy metal precipitates is favoured at acid pH, which is the favourable pH for soil or sludge decontamination, whereas an alkaline pH is favourable for precipitation of heavy metal ions to decontaminate wastewater.

An example of the formation of precipitates or soluble compounds at different redox states occur in ferrous and manganese compounds. Under aerobic conditions Fe(III) (ferric iron) and Mn(IV) ions form insoluble Fe(III) and Mn(IV) oxides or hydroxides, but under anaerobic conditions they form soluble Fe(II) compounds (ferrous iron) or Mn(II) compounds. Most organisms that grow with energy conserved during reduction of Fe(III) or Mn(IV) are members of the *Geobacteriaceae*. Transfer of electrons from the terminal reductase, localised in the outer membrane or at the cell surface, to the insoluble Fe(III) or Mn(IV) oxides outside the cells can proceed either in a direct way (contact between the oxides and the cells) or by 'soluble electron shuttles' (e.g., by humic substances) between the metal-reducing microorganism and the mineral (Lloyd, 2003). Since Fe is an essential element for microorganisms, aerobic bacteria must excrete siderophores, which bind Fe^{3+} to their phenolate or hydroxamate moiety and supply the cells with soluble Fe^{2+}. To accelerate Fe^{3+} reduction in biotechnological processes, the chelator nitrilotriacetic acid can be added.

Metals as electron acceptors for anaerobic respiration.

Reaction	*Microorganism*	*Reference*
$2\ Fe^{3+} + H_2 \leftrightarrow 2\ Fe^{2+} + 2\ H^+$	*Geobacter metallireducens*	Lovley and Lonergan (1990)
	Pelobacter carbinolicus	Lovley *et al.* (1995)
$Mn^{4+} + H_2 \leftrightarrow Mn^{2+} + 2\ H^+$	*Geobacter metallireducens*	Lovley (1991)
	mixed culture	Langenhoff *et al.* (1997)
$2\ Cr^{6+} + 3\ H_2 \leftrightarrow 2\ Cr^{3+} + 6\ H^+$	*Desulphovibrio vulgaris*	Lovley and Phillips (1994)
	Bacillus strain QC1-2	Campos *et al.* (1995)

Contd...

Reaction	*Microorganism*	*Reference*
$Se^{6+} + H_2 \leftrightarrow Se^{4+} + 2\ H^+$	*Thauera selenatis* strains SES-1; SES-3	Macy *et al.* (1993)
$Se^{6+} + 3\ H_2 \leftrightarrow Se^{\circ} + 6\ H^+$		Oremland *et al.* (1989)
$Te^{4+} + 2\ H_2 \leftrightarrow Te^{\circ} + 4\ H^+$	*Schizosaccharomyces pombe*	Smith (1974)
$Pb^{2+} + H_2 \leftrightarrow Pb^{\circ} + 2\ H^+$	*Pseudomonas maltophila*	Lovley (1995)
$As^{5+} + H_2 \leftrightarrow As^{3+} + 2\ H^+$	*Geospirillum arsenophilus*	Ahmann *et al.* (1994)
$Hg^{2+} + H_2 \leftrightarrow Hg^{\circ} + 2\ H^+$	*Escherichia coli* Thiobacillus ferrooxidans	Robinson and Tuovinen (1984)
$U^{6+} + H_2 \leftrightarrow U^{4+} + 2\ H^+$	*Shewanella putrefaciens*	Lovley *et al.* (1991)

In addition to their use in synthesis of cell components (e.g., cytochromes, ferredoxin, etc.), Fe^{2+} salts can be electron donors for nitrate reduction (Straub *et al.*, 1996).

The reduction of Hg^{2+} to metallic Hg by *Escherichia coli* or *Thiobacillus ferrooxidans* facilitates Hg separation and prevents methylation reactions under aerobic or sulphate-reducing conditions. The mechanism includes non-enzymatic transfer of methyl groups from methylcobalamin to Hg^{2+} to form methyl mercury or dimethyl mercury, which are both neurotoxins and become enriched in the food chain. Selenium and arsenic can be transformed microbiologically by methylation to dimethyl selenide or to di-ortrimethyl arsine in a volatile form. Methylated arsenic compounds are less toxic than non-methylated arsenic compounds (White *et al.*, 1997).

Changing the redox state of natural and anthropogenic radionuclides by metal-reducing microorganisms offers a possibility to control their solubility and mobility by converting, e.g., U^{6+} to U^{4+}, Pu^{5+} to Pu^{4+}, or Np^{5+} to Np^{4+}. The tetravalent metals can be removed by chelators (e.g., EDTA) or by immobilisation onto biomass from the contaminated environment (Lloyd, 2003).

In contrast to bacteria, fungi are capable of leaching soluble as well as insoluble metal salts, because they excrete organic acids such as citric acid, fumaric acid, lactic acid, gluconic acid, oxalic acid, or malic acid, which dissolve metal salts and form complexes with the metal ions. The leaching efficiency depends on the soil microflora.

Some soil microorganisms seem to be able to degrade the carbon skeleton of the metal–organic complex and thus immobilise the metal ions again (Brynhildsen and Rosswall, 1997).

Mixed bacterial cultures or *Wolinella succinogenes* use perchlorate or chlorate as electron acceptors for respiration (Wallace *et al.*, 1996; van Ginkel *et al.*, 1995) and thus detoxify these chemicals.

Sulphate Reduction and Metal Ion Precipitation

Sulphate-reducing bacteria are biotechnologically relevant to sulphate removal or heavy metal precipitation in wastewater or waste and to the elimination of SO_2 during offgas purification. An overview of applications of sulphate-reducing microorganisms in environmental biotechnology is given by Lens *et al.* (2002). Sulphate is the terminal electron acceptor and is reduced to sulphide, with reducing equivalents derived from the degradation of lactic acid or many other organic compounds (Widdel, 1988).

Alternatively, some sulphate reducers can also use molecular hydrogen. Sulphate reducers gain energy in an anaerobic electron transport chain (Hansen, 1994), leading to sulphide, a weak dibasic acid, which dissociates according to Eq. 35.

$$H_2S \xleftrightarrow{K_1} H^+ + HS^- \xleftrightarrow{K_2} 2\ H^+ + S^{2-}$$

$$K_1 = 1.02 \times 10^{-7},\ K_2 = 1.3 \times 10^{-13}\ (25°C) \quad (35)$$

The total dissociation is described by Eq. 36:

$$K = = K_1K_2 = 1.3 \times 10^{-20} \quad (36)$$

For precipitation of heavy metal ions, sulphide ions are necessary (Eq. 37):

$$Me^{2+} + S^{2-} \leftrightarrow MeS\ \text{(or)}\ 2\ Me^+ + S^{2-} \leftrightarrow Me_2S \quad (37)$$

The concentration of sulphide is pH-dependent. At acid pH only those metal sulphides of very low solubility can be precipitated. Thus, at acid pH, HgS, As_2S_3, CdS, CuS, and PbS form precipitates, whereas at a more alkaline pH, ZnS, FeS, NiS, and MnS form precipitates. Al_2S_3 and Cr_2S_3 are water soluble and cannot be removed by precipitation or sedimentation. Zinc removal from zinc-contaminated groundwater by microbial sulphate reduction and zinc sulphide precipitation in a 9-m^3 sludge blanket reactor was demonstrated and has been transferred to a full scale reactor of 1800 m^3 (White and Gadd, 1996).

Aerobic and Anaerobic Degradation of Xenobiotics

Except for pesticides and insecticides, which are widely used in horticulture and farming and small amounts of which are washed into the groundwater, many xenobiotics are spilled in the environment, by spot contamination or contamination of large areas. Biological remediation of soil and groundwater contaminated with gasoline (e.g., Yerushalmi and Guiot, 1998) and the volatile fractions of diesel oil (e.g., Greiff *et al.*, 1998), even at low temperatures (Margesin and Schinner, 1998), seems to be no problem. Results of many laboratory and field studies with a variety of substances are available (e.g., Arendt *et al.*, 1995; Hinchee *et al.*, 1995a, b; Kreysa and Wiesner, 1996). Except for studies in complex field or wastewater environments, many xenobiotics can be degraded by aerobic or anaerobic pure or defined mixed cultures. These include monoaromatic and polyaromatic substances with or without chloro substituents. Whereas the two-ring compound naphthalene is relatively easily degradable by *Pseudomonas* sp. and *Rhodococcus* sp., the four-ring compounds fluoranthene and pyrene are much less degradable by *Rhodococcus* sp. (Bouchez *et al.*, 1996). Aromatic compounds such as phenol and benzoic acid are degradable by mixed consortia at rates up to 1g L^{-1} d^{-1} in either the presence or absence of oxygen (Morsen and Rehm, 1987; Knoll and Winter, 1987; Kobayashi *et al.*, 1989). A potent population is required, which can be obtained by preincubation.

Even PCP is biodegradable by a methanogenic mixed culture from UASB granules (Juteau *et al.*, 1995; Wu *et al.*, 1993; Kennes *et al.*, 1996). Some pure cultures and mixed cultures are capable of dechlorinating aromatic or aliphatic compounds. As for the aerobic and anaerobic degradation of phenol, the rates of dechlorination of 2-chlorophenol under aerobic and anaerobic conditions are the same order of magnitude, 102 and 128 mg L^{-1} d^{-1}, respectively (Kafkewitz *et al.*, 1996; Dietrich and Winter, 1990). A pure culture of *Pseudomonas pickettii* was used for aerobic dechlorination, but for anaerobic degradation a mixed culture was used.

The main problem in degradation of xenobiotic compounds in wastewater is a too short residence time in the reactors, which does not allow selection or adaptation of bacteria for dechlorination or degradation. Only if a permanent pollution is prevalent, the degradation potential may develop with time. Alternatively, biofilm reactors should be used to enrich dechlorinating and xenobiotic-degrading bacteria.

Table: Aerobic and anaerobic degradation or dechlorination of xenobiotics

Reaction Substance	*Rate (mg L–1 d–1)*	*Microorganisms*	*Reference*
Aerobic degradation			
2-Hydroxy-benzothiazole	138	*Rhodococcus rhodochrous*	De Wever *et al.* (1997)
Naphthalene	57	*Pseudomonas* *Rhodococcus* sp.	sp. Bouchez *et al.* (1996)
Fluoranthene	6.6	*Rhodococcus* sp.	Bouchez *et al.* (1996)
Pyrene	6.6	*Rhodococcus* sp.	Bouchez *et al.* (1996)
Pyrene	0.56	*Mycobacterium flavescens*	Dean-Ross and Cerniglia (1996)
Toluene	57	*Pseudomonas putida*	Heald and Jenkins (1996)
Phenol	188	*Bacillus* sp.	A2 Mutzel *et al.* (1996)
Phenol	1000	mixed immobilised culture	Morsen and Rehm (1987)

Contd...

Reaction Substance	*Rate (mg L–1 d–1)*	*Microorganisms*	*Reference*
Cresol (*o, m, v*)	259	mixed immobilised culture	Morsen and Rehm (1987)
2,4-Diphenoxy-acetic acid	33	*Alcaligenes eutrophusa* JMP134 (pJP4)b	Valenzuela *et al.* (1997)
2,4,6-Trichlorophenol	15	JMP134 (pJP4)b	Valenzuela *et al.* (1997)
Anaerobic Degradation			
Pentachlorophenol	107	methanogenic mixed culture, fixed film reactor	Juteau *et al.* (1995) (PCP)
PCP	90	methanogenic mixed culture UASB	Wu *et al.* (1993)
PCP	4.4	methanogenic granules	Kennes *et al.* (1996)
PCP	22.7	methanogenic mixed culture	Juteau *et al.* (1995)
Benzene	0.029	sulphate-reducing mixed culture	Edwards and Gribic´-Galic (1992)
Phenol	1000	methanogenic mixed culture	Knoll and Winter (1987)
Phenol	31	syntrophic mixed culture	Knoll and Winter (1989)
Phenol	200	syntrophic mixed culture	Kobayashi *et al.* (1989)
Benzoic acid	600	syntrophic culture	Kobayashi *et al.* (1989)
Toluene	0.1–1.5	sulphate-reducing mixed culture	Edwards *et al.* (1992)
Toluene	4.6	methanogenic mixed culture	Edwards and Gribic-Galic (1994)
Xylene	0.1–1.5	sulphate-reducing mixed culture	Edwards *et al.* (1992)
Xylene	5.3	methanogenic mixed culture	Edwards and Gribic-Galic (1992)
Aerobic Dechlorination			
2-Chlorophenol	102	*Pseudomonas pickettii*	Kafkewitz *et al.* (1996)
4-Chlorophenol	41	*Pseudomonas pickettii*	Kafkewitz *et al.* (1996)
1,3-Dichloro-2-propanol	671	*Pseudomonas pickettii*	Kafkewitz *et al.* (1996)
Tetrachloroethene	35.8	totalc anaerobic mixed culture	Wu *et al.* (1995)
Anaerobic Dechlorination			
2,6-Dichlorophenol	38.4	methanogenic mixed culture	Dietrich and Winter (1990)

Contd...

Reaction Substance	*Rate (mg L–1 d–1)*	*Microorganisms*	*Reference*
4-Chlorophenol	0.43	sulphate-reducing consortium	Haggblom (1998)
2-Chlorophenol	128 mixed culture	methanogenic (1990)	Dietrich and Winter
2-Chlorophenol	1.66	anaerobic mixed culture	Kuo and Sharak Genthner (1996)
3-Chlorobenzoate	6.08 culture	anaerobic mixed Genthner (1996)	Kuo and Sharak
3-Chloro-4-hydroxy-benzoate	29.9	*Desulphitobacterium chlororespirans*	Sanford *et al.* (1996)
Polychlorinated biphenyls (PCBs)			
2,3,4,5,6-CB[d]	0.24	total methanogenic granules	Natarajan *et al.* (1996)
2,3,4,5-CB	0.39 3,5-CBc	anaerobic sediments	Berkaw *et al.* (1996)
2,3,4,6-CB	13.3 tri-CBc	anaerobic sediment	Wu *et al.* (1996)
Tetrachloroethene	6.13	methanogenic granules (1997)	Christiansen *et al.*
DCB, TCB, TeCB[e]	1.24	methanogenic mixed culture	Middeldorp et at. (1997)

a. Alcaligenes eutrophus = Ralstonia eutropha.

b. pJP4= 2,4-dicholorophenoxy acetic acid-degrading plasmid.

c. Dehalogenation product: total dehalogenation or partial dehalogenation to the corresponding dehalogenation product.

d. 2,3,4,5,6-CB= 2,3,4,5,6-chlorinated biphenyls.

e. DCB= dichlorobenzene, TCB=trichlorobenezene, TeCB = tetrachlorobenzene.

Bioaugmentation in Wastewater Treatment Plants for Degradation of Xenobiotics

In biotechnology and pharmacology, mutants of bacteria or fungi or genetically engineered organisms are widely used in the production of citric acid, gluconic acid, ascorbate, and pharmaceuticals such as penicillin, insulin, and blood coagulation factors.

Bacteria and fungi have also been adapted or genetically

transformed for soil remediation (Atlas, 1981; Margesin and Schinner, 1997, Korda *et al.*, 1997; Megharai *et al.*, 1997). For wastewater and sludge stabilisation however, successful use of genetically modified bacteria or of bacteria that can serve as donors for plasmids encoding degradative enzymes has been rather rare (van Limbergen *et al.*, 1998). Usually, natural selection of the most suitable microorganisms from a complex flora, simply by adapting process parameters, is used. The limited reports on successful bioengineering for wastewater treatment may result from any of several factors:

- The plasmids were unstable or the genes are not expressed in the new environment.
- Inoculated strains did not survive or, if they survived, metabolic activity was too low for successful competition with autochthonous strains.
- Inoculated strains, serving as a gene source, survived, but other strains were not competent for gene transfer.
- Wastewaters normally contain a complex spectrum of carbon sources that are better than xenobiotics, and so organisms do not express genes for degrading xenobiotics.

To increase the survival potential in wastewater, a selected flora should have the desired degradative ability, tolerance to cocontaminants, and a natural spatial and temporary abundance (van der Gast *et al.*, 2003). Another possibility would be to add organisms containing plasmids having a broad host range, permitting conjugation and DNA exchange between different species or genera of bacteria.

To enhance degradation of organic compounds in activated sludge or by a biofilm, selected specialised bacteria, genetically modified bacteria, or bacteria as plasmid donors for degradative pathways can be added (McClure *et al.*, 1991; Frank *et al.*, 1996). Plasmid exchange from donor to recipient cells would be advantageous to the recipient bacteria, because the plasmids harbour genes whose products function in pathways for xenobiotic degradation, which would thus extend the substrate spectrum of the recipients. The transfer of naturally occurring mercury-resistance plasmids between *Pseudomonas* strains in biofilms was shown to occur rapidly (Bale *et al.*, 1988).

Selvaratnam *et al.* (1997) added a phenol-degrading strain of *Pseudomonas putida* to an activated sludge SBR reactor, which had removed 170 mg L^{-1} of phenol before being augmented with the *Pseudomonas putida* strain. Whereas the original phenoldegrading activity was partially lost in the non-augmented reactor upon further operation, the augmented reactor almost completely degraded the phenol. A more convincing approach would have been to use non-phenol–degrading activated sludge for this experiment, although the survival of the catabolic plasmid *dmpN* of *Pseudomonas putida* and its expression in the reactor biomass was demonstrated for 44 d by molecular biology techniques under steady-state conditions in the laboratory.

Successful bioaugmentation experiments in an upflow anaerobic sludge blanket reactor were reported by Ahring *et al.* (1992), who introduced a suspension of a pure culture of *Desulphomonile tiedjei* or a three-member consortium into an UASB reactor.

They observed a rapid increase in the dehalogenation of 3-chlorobenzoate, whereas non-amended parallel incubations had no dehalogenating activity. Even after several months at 0.5-d hydraulic residence time, which was shorter than the generation time of *Desulphomonile tiedjei*, dehalogenating activity was still observed, and *Desulphomonile tiedjei* could be found within the biofilm by the use of antibody probes.

More recently, an UASB reactor was supplemented with *Dehalospirillum multivorans* to improve its dehalogenating activity (Horber *et al.*, 1998). In contrast, Margesin and Schinner (1998) found, for biological decontamination of fuel-contaminated wastewater, that stimulation of the autochthonous flora by adding a mineral mix enhanced biodegradation to a larger extent than bioaugmentation with a cold-adapted mixed inoculum containing *Pseudomonas* sp. and *Arthrobacter* sp. Bioaugmentation with a complex inoculum, which naturally occurs in metalwork fluids and contains strains of *Clavibacter* sp., *Rhodococcus* sp., *Methylobacterium* sp. and *Pseudomonas* sp. is very effective in degrading COD in wastewater. The augmented consortium degraded 67 per cent of the COD (48 g L^{-1}) and was therefore 50 per cent–60 per cent more effective than the indigenous flora. *In situ* analysis showed that 100 h after the introduction, the augmented consortium constituted more than 90 per cent of the population (van der Gast *et al.*, 2003).

Degradation of many xenobiotics in anaerobic or aerobic pure cultures or complex ecosystems has been demonstrated. Whereas for 'intrinsic sanitation' of polluted soil, the time is not limited, so long as the pollutants are adsorbed tightly to the soil matrix, degradation of xenobiotics in wastewater must be completed during the residence time in the treatment system. Thus, merely the presence of a degradation potential within a wastewater ecosystem is not sufficient, but degradation rates must be high enough, and degradation must be faster than the residence time of bacteria in suspended systems. Supplementation with microorganisms as a potential tool to increase the degradation speed or to increase the degradation potential in wastewater cannot in general be considered a state-of-the-art procedure yet.

Aerobic Glucose Respiration

Only about 5 per cent of the fermentable substrate is used for cell growth (surplus sludge formation) in anaerobic reactors, whereas 95 per cent is converted to methane and CO_2, and most of the energy of the substrates is conserved in the fermentation products.

Mass and Energy Balance for Aerobic Glucose Respiration and Sewage Sludge Stabilisation in most writings of microbiology, respiration of organic matter is explained with glucose used as a model substance. Except for an exact reaction stoichiometry of the oxidative metabolism, mass and energy dissipation, if mentioned at all, are not quantified. Both parameters are, however, very important for activated sludge treatment plants. The surplus sludge formed during wastewater stabilisation requires further treatment, causes disposal costs, and – in the long run – may be an environmental risk, and heat evolution during unevenly high-loaded aerobic treatment may shift the population towards more thermotolerant or thermophilic species and thus, at least for some time, may decrease the process efficiency.

Biological wastewater treatment is among the most important biotechnological applications and, as drivers of the key processes, microorganisms are central to its success. Therefore, the study of wastewater microorganisms has obvious applied significance; however, the importance of wastewater treatment reactors as model systems for microbial ecology is often overlooked. Modern molecular techniques, including environmental genomics, have identified unexpected microbial key players for nutrient removal and sludge bulking and/or foaming, and provided many exciting insights into the diversity, functions and niche differentiations of these predominantly uncultivated microorganisms. It is now time for wastewater microbiology to be recognised as a mature and dynamic discipline in its own right, offering much towards a deeper understanding of life in complex microbial communities. Here, we consider selected key findings to illustrate the past and future roles of molecular ecophysiology and genomics in the development of wastewater microbiology as an important subdiscipline of microbial ecology.

Primary and secondary treatment remove the majority of BOD and Suspended Solids found in wastewaters. However, in an increasing number of cases this level of treatment has proved to be insufficient to protect the receiving waters or to provide reusable water for industrial and/or domestic recycle. Thus, additional treatment steps have been added to wastewater treatment plants to provide for further organic and solids removals or to provide for removal of nutrients and/or toxic materials.

For the purposes of this lesson, Advanced Wastewater Treatment will be defined as: any process designed to produce an effluent of higher quality than normally achieved by secondary treatment processes or containing unit operations not normally found in Secondary Treatment. The above definition is intentionally very broad and encompasses almost all unit operations not commonly found in wastewater treatment today.

Types of Advanced Wastewater Treatment

Advanced Wastewater Treatment may be broken into three major categories by the type of process flow scheme utilised:

1. Tertiary Treatment
2. Physical-Chemical Treatment
3. Combined Biological-Physical Treatment

Tertiary treatment may be defined as any treatment process in which unit operations are added to the flow scheme following conventional secondary treatment. Additions to conventional secondary treatment could be as simple as the addition of a filter for suspended solids removal or as complex as the addition of many unit processes for organic, suspended solids, nitrogen and phosphorous removal. Physical-chemical treatment is defined as a treatment process in which biological and physical-chemical processes are intermixed to achieve the desired effluent. Combined biological-physical-chemical treatment is differentiated from tertiary treatment in that in tertiary treatment any unit processes are added after conventional biological treatment, while in combined treatment, biological and physical-chemical treatment are mixed.

Another way to classify advanced wastewater treatment is to differentiate on the basis of desired treatment goals. Advanced wastewater treatment is used for:

4. Additional organic and suspended solids removal.
5. Removal of nitrogenous oxygen demand (NOD).
6. Nutrient removal.
7. Removal of toxic materials.

In many, if not most instances today, conventional secondary treatment gives adequate BOD and suspended solids (SS) removals. Why, then, is additional organic and suspended solids removal by advanced treatment necessary? There are a number of good answers to this question.

8. Advanced wastewater treatment plant effluents may be recycled directly or indirectly to increase the available domestic water supply.
9. Advanced wastewater treatment effluents may be used for industrial process or cooling water supplies.
10. Some receiving waters are not capable of withstanding the pollutional loads from the discharge of secondary effluents.
11. Secondary treatment does not remove as much of the organic pollution in wastewater as may be assumed.

The first three reasons for additional organic removal through advanced wastewater treatment are simple. The fourth requires some explanation. The performance of secondary treatment plants is almost always measured in terms of BOD and SS removals. A well designed and operated secondary plant will remove from 85 to 95 per cent of the influent BOD and SS. However, the BOD test does not measure all of the organic material present in the wastewater.

An average secondary effluent may have a BOD of 20 mg/L and a COD of 60 to 100 mg/L. The average secondary plant removes approximately 65 per cent of the influent COD. Thus, when high quality effluents are required, additional organic removals must be accomplished. In addition to the organic materials remaining in most secondary effluents, there is an additional oxygen demand resulting from the nitrogen present in the wastewater.

In wastewaters, much of the nitrogen is found in the form of ammonia. When secondary treatment is used, a great deal of this ammonia is discharged in the effluent. Bacteria can utilise this ammonia as an energy source and convert ammonia to nitrite and nitrate.

$$NH_3 + O_2 + \text{Bacteria} \rightarrow NO_2 + O_2 + \text{Bacteria} \rightarrow NO_3$$

Another reason for advanced wastewater treatment may be to remove nutrients contained in discharges from secondary treatment plants. The effluents from secondary treatment plants contain both nitrogen (N) and phosphorous (P). N and P are ingredients in all fertilizers. When excess amounts of N and P are discharged, plant growth in the receiving waters may be accelerated. Algae growth may be stimulated causing blooms which are toxic to fish life as well as aesthetically unpleasing. Fixed plant growth may also be accelerated causing the eventual process of a lake becoming a swamp to be speeded up. Therefore, it has become necessary to remove nitrogen and phosphorous prior to discharge in some cases.

Toxic materials, both organic and inorganic are discharged into many sewage collections systems. When these materials are present in sufficient quantities to be toxic to bacteria, it will be necessary to remove them prior to biological treatment. In other cases, it is necessary to remove even small amounts of these materials prior to discharge to protect receiving waters or drinking water supplies. Thus, advanced wastewater treatment processes have been used in cases where conventional secondary treatment was not possible due to materials toxic to bacteria entering the plant as well as in cases where even trace amounts of toxic materials were unacceptable in plant effluents.

Nitrification

Biological nitrification may be used to prevent oxygen depletion from nitrogenous demand (NOD) in the receiving waters. Nitrification is simply the conversion of ammonia to nitrate in the treatment plant rather than in the receiving water. Nitrification may be carried out in the same tank as BOD removal or in a separate stage. Nitrification may be carried out either in activated sludge flocs or in fixed films.

Regardless of the particular scheme chosen, the same basic requirements for nitrification must be maintained:

1. Oxygen level
2. Loading rates
3. Solids retention time
4. Alkalinity
5. pH
6. Freedom from toxic materials
7. Temperature

Sufficient oxygen must be available for nitrification to occur. Approximately 4.5 pounds of dissolved oxygen are required for the conversion of 1 pound of ammonia to nitrate. Dissolved oxygen sufficient to satisfy the remaining BOD is also required. In activated sludge plants the mixing requirement of the basins must also be considered. Generally, dissolved oxygen levels of approximately 2 - 3 mg/L are recommended for nitrification. The bacteria responsible for nitrification reproduce at a much slower rate than those responsible for BOD removal. Thus, the danger always exists for the "wash out" of the nitrifying organisms. That is, unless the nitrifying bacteria reproduce at the same or greater rate than they are removed from the system (by waste sludge) then the population of bacteria will be insufficient to carry out nitrification. For this reason, nitrification systems are operated at higher return sludge rates than conventional secondary treatment. The amount of sludge to be wasted is significantly less than from a conventional activated sludge system.

Nitrification systems are sensitive to pH variation. Optimum pH has been found to be approximately 7.8 to 9.0. Reductions in nitrification have been found outside this range. Alkalinity is also destroyed during nitrification. Theoretically, 7.2 pounds of alkalinity are destroyed in converting 1 pound of ammonia to nitrate. In low alkalinity wastewaters, Quick lime (CaO) or $Ca(OH)_2$ is often used to provide alkalinity and pH control.

Generally, the influent BOD to nitrification systems has not been found to effect performance. However, sufficient oxygen must be provided for the organic demand and organic shock loads must be avoided.

Biological Denitrification

Biological nitrification satisfies the nitrogenous oxygen demand by converting NH_3 to NO_3. In some applications, such as discharge into enclosed bodies of water or recycle to water supplies, nitrification may not be sufficient. When nitrogen removal is required, one of the available methods is to follow biological nitrification with biological denitrification.

Denitrification is accomplished under anaerobic or near anaerobic conditions by bacteria commonly found in wastewater. Nitrates are removed by two mechanisms:

(1) Conversion of NO_3 to N_2 gas by bacterial metabolism, and

(2) Conversion of NO_3 to nitrogen contained in cell mass which may be removed by settling.

In order for denitrification to occur, a carbon source must be available. Most commonly, methanol is used. The methanol must be added in sufficient quantity to provide for cell growth and to consume any dissolved oxygen which may be carried into the denitrification reactor.

Usually 3 to 4 pounds of methanol per pound of nitrate are required. Careful control of methanol feed is necessary to prevent waste of chemicals. In addition, if excess methanol is fed to the system, unused methanol will be carried out in the effluent causing excessive BOD.

Denitrification may be carried out in either a mixed slurry reactor or in fixed bed reactors. Denitrification filters carry out both denitrification and filtration in the same unit. Mixed slurry systems consist of a denitrification reactor, reaeration basin and clarifiers. Reaeration prior to clarification is required to free the sludge from trapped bubbles of nitrogen gas.

Denitrifying bacteria grow very slowly and are extremely sensitive to temperature. Denitrification rates have been shown to increase fivefold when the temperature is increased from 10°C to 20°C. Thus, operating parameters such as sludge age and retention time must be varied with temperature.

The pH in denitrification systems must be carefully controlled. The optimum pH is from 6.0 to 8.0.

Denitrification is a very sensitive and difficult process to operate. Little full scale operational experience is available. Constant monitoring of pH, methanol feed and temperature is essential to successful operation.

Filtration

Granular media filtration to remove those suspended and colloidal solids which are carried over from previous unit processes is a common unit process in advanced wastewater treatment. Effluents of less than 10 mg/L BOD and 5 mg/L suspended solids are not uncommon for effluents from biological treatment processes after filtration.

Gravity filters similar to rapid sand filters are sometimes used. Often a combination of filter medias, such as anthracite coal and sand are used to provide coarse to fine filtration as the water passes through the filter. The water passes through the filter media and support gravel and is then collected by the underdrain system. As filtration proceeds, the headloss through the filter increases until it reaches an unacceptable level or until solids breakthrough occurs and the effluent becomes unacceptable. When either the headloss becomes excessive or solids breakthrough occurs, the filter is backwashed.

Gravity filters are generally run at 1.5 to 2.5 gpm per square foot. Pressure filters are used to obtain filter rates up to 6 gpm per square foot. Ideally, filters are designed to have the solids in the effluent and the headloss reach their allowable levels at the same time.

Advanced wastewater treatment can be used to achieve any level of treatment desired. Advanced wastewater treatment plants in Lake Tahoe, California and Windhoek, South Africa have been achieving drinking water quality since 1968. Advanced wastewater treatment plants utilise sophisticated processes and equipment. They are relatively expensive to run and operating costs as well as effluent quality are sensitive to the quality of operation.

Septic Tank Sludge (On-site Treatment)

On-site treatment is used when domestic sewage is not routed to a sewage system. The simplest method of containment of faecal matter, and one that is still used by hundreds of millions of people throughout the world, is the use of dry latrines – i.e., a simple hole in the ground that is covered once full. A more evolved form of on-site treatment is the use of a septic tank, which is the method generally used in industrialised countries to treat domestic sewage from isolated residences. Typical domestic wastewater from a North American home contains some 450 mg/L of suspended solids and an organic load equivalent to 400 mg/L of BOD.

Wastewater treatment in septic tank systems occurs in two stages. First, effluent from the home enters an underground tank that holds between 1500 and 5000 L; the average retention time in the tank is 2-3 days. The partially treated effluent flows from the tank and seeps into the ground through a system of perforated pipes. The soil is a natural filter and has the capacity to complete the treatment that began in the tank.

Septic tanks are underground and are generally constructed of concrete, plastic, or fibreglass. They generally have two chambers separated by a wall that allows the wastewater to flow easily from the first chamber into the second. Solids that are denser than water sink to the bottom, whereas oil, grease, and low-density solids float on the surface, forming the scum layer or crust. The wastewater flows continually, whereas the sediments must be regularly pumped (every few years) to prevent overload of the system.

Microbial Dynamics of Septic Tanks

As with municipal wastewater treatment, the fresh organic matter that enters the septic tank is metabolised by various microorganisms. A first group of facultative anaerobic and aerobic bacteria converts the complex organic compounds into simple substances such as monosaccharides, amino acids, and volatile fatty acids. A second group of strict anaerobic bacteria then converts the initial metabolites into molecules such as CO_2, hydrogen (H_2), CH_4, and H_2S. The formation of CH_4 is uncommon in Canada because methanogenic bacteria grow very slowly below 20°C. The main gases that are produced break away from the sediments in the form of bubbles and, as they rise to the surface, entrain particles of sludge to form part of the scum layer.

Mineralisation of organic matter generally occurs after a retention time of one year in the septic tank. The production of CH_4, if possible, does not begin until the second year of operation without pumping. Pumping the septic tank every two years therefore will prevent methanisation. The regular pumping of the septic tank is required to prevent the sludge from overflowing into the second chamber, then into the tile field; an excess of total suspended solids in septic tank effluent could plug the pipes that carry the effluent to the tile field.

The substances formed by anaerobic decomposition are responsible for the unpleasant and penetrating odour of the sludge. The odours are caused in part by gases, such as H_2S, but also by volatile fatty acids and compounds such as indole and skatole, which are responsible for faecal odour.

Quantity of Sludge Produced

In general, the rate of annual sludge accumulation gradually declines over time. After the first year of septic tank operation, sludge and scum production ranges from 80 to 180 L per person, falling to 40-100 L after three years of operation. The tank's contents should be completely pumped out to promote a higher rate of sludge accumulation the following year. The decline in the rate of sludge accumulation over time is normal with anaerobic digestion, a process that reduces microbial biomass.

A low annual temperature promotes the accumulation of sludge owing to the slowdown in microbial activity responsible for the degradation of the organic matter. Below 10°C, bacterial metabolism is virtually inhibited, which results in a considerable increase in sediments during the cold season. Discharges of large quantities of organic matter promote an increase in sludge and reduced water flow, which will tend to result in poor operation of the septic tank.

Pollutant Load of Septic Tank Sludge

Septic tank sludge consists of scum (composed of greases and floating solids), sediments, and the water removed from the septic tank during pumping. Apart from various non-decomposable solid materials (sand, pieces of plastic or metal), the sludge is essentially composed of a dark brown to black mass.

Septic tank sludge contains very high levels of contaminants, as can be seen from the following figures:

- BOD, 5000-10 000 mg/L;
- total suspended solids, 20 000-40 000 mg/L;
- total nitrogen, 100-1000 mg/L; and
- total phosphorus, 30-300 mg/L.

According to the available data, septic tank sludge contains no hazardous organic compounds, such as chlorinated substances (e.g., PCBs) or PAHs. However, surfactants from detergents, such as nonylphenol or alkyl benzene, may be present, as well as phthalates, which are used in the production of adhesives, glues, and lubricants.

During normal operation, the concentrations of pollutants in treated wastewater leaving the septic tank are reduced: BOD, 30-250 mg/L; total suspended solids, 30-200 mg/L; total nitrogen, 10-300 mg/L; and total phosphorus, 1-20 mg/L. This effluent is carried to a tile field, where the concentration of the pollutants is further reduced by microorganisms in the soil.

The sludge recovered during septic tank pumping has a high pollutant load, and its management is problematic because it contains several groups of potentially toxic or pathogenic substances.

Heavy Metals

Septic tank sludge contains low concentrations of heavy metals. The main sources of metals are domestic sanitary products and metal pipes. The primary metals found in septic tank sludge are as follows:

- iron, roughly 200 mg/L;
- aluminium, 50-250 mg/L;
- zinc, roughly 35 mg/L; and
- copper, roughly 10 mg/L.

These concentrations are usually below the values recommended for the use of sewage sludge in agriculture (iron, 1000 mg/L; zinc, 500 mg/L; copper, 100 mg/L), with the exception of aluminium, the maximum concentration of which should not exceed 100 mg/L.

Microorganisms

The microbial flora of septic tank sludge is similar to that of primary sludge from a municipal wastewater treatment plant. The type and number of the key organisms identified are as follows (per 100 mL):

- total coliforms, 10^7-10^9;
- faecal coliforms, 10^6-10^8;
- faecal streptococci, 10^6-10^7;
- sporulating anaerobic bacteria (primarily *Clostridium* and Bacteroides), approximately 10^5; and
- salmonellae, 1-100.

Septic tank sludge also contains various parasites, such as

Ascaris and *Trichuris*, as well as many viruses, such as rotaviruses, the Norwalk virus, and enteric adenoviruses.

Special caution must therefore be exercised in the handling and management of septic tank sludge because, like municipal sewage sludge, it can cause various gastrointestinal infections.

Sludge Management

The reclamation of septic tank sludge is desirable. However, if there are no other options, the disposal of septic tank sludge at sanitary landfills is acceptable. There are four main management methods:

1. impoundment;
2. dewatering;
3. agricultural reclamation; and
4. discharge into municipal sewage systems.

In addition, composting is a promising technique that is currently being developed.

Impoundment

Impoundment is a form of anaerobic stabilisation that allows inert settleable solids to separate from the liquid and sink to the bottom. The sludge is poured into lagoons, which are usually installed at sanitary landfill sites. The liquid part of the sludge seeps into the soil and is treated by natural microflora. This practice is inexpensive, but there is a risk of contamination of the water table by the lagoons. This type of management is coming under increasing criticism and is not a long-term solution, particularly in terms of public health risks.

Dewatering

Dewatering by fixed or mobile equipment produces a relatively solid "pelletable" sludge, which can be sent to a sanitary landfill for disposal. The filtrate (liquid fraction removed from the sludge) must, however, be sent to an on-site treatment system or to a municipal wastewater treatment system.

Agricultural Reclamation

The use of septic sludge in agriculture is possible, but there are two factors that pose obstacles to this practice: the cost of transporting small volumes from each septic tank; and the need to stabilise the sludge in order to destroy the pathogens and to mineralise the organic matter. Moreover, the sludge cannot be land-applied at just any time, which means that temporary storage structures must be available.

The application of raw sludge to agricultural or forest land is not recommended because the organic matter is not sufficiently broken down to meet the fertilization needs. Moreover, the presence of pathogenic microorganisms in fresh sludge must also be considered. Studies have shown that the composting of septic tank sludge is technically feasible. In addition, the resulting compost meets agronomic standards associated with available nutrients (particularly nitrogen and phosphorus), organic amendment, and the number of pathogenic microorganisms.

Municipal Sewage Systems

Septic sludge can be discharged directly into municipal sewage systems if volumes are low relative to the capacity of the treatment plant. Treatment plants are generally not designed to treat water with high concentrations of total suspended solids and a very high BOD. This type of management is not recommended.

It is very difficult to quantify the health risks associated with the management of septic sludge because no epidemiological follow-up is carried out on persons handling septic tank sludge on a daily basis.

8

Utilisation of Solid Wastes

Building materials account for the major component of the construction cost depending upon the locations, namely rural, small town, medium town, cities and metropolitan context as well as nature of materials being used. Building materials contribute to 60 to 70 per cent of the cost of construction. The prominent use of traditional and conventional construction materials like steel, cement, burnt bricks and timber clearly manifests a interesting trend in construction cost over the years. The annual increase in cost of construction has always been over the annual inflation. With the inflation about 9 to 11 per cent, the building construction cost have jumped to 15 to 18 per cent and with the recent hike in cost of steel and cement, it is bound to show higher trends.

The shortage of building materials has been reckoned as cement 17 million tonnes, steel 2.7 million tonnes, bricks 54898 millions, aggregate 286 million cubic metres and timber 13.34 million cubic metres at present. Data show that there is a considerable amount of shortage of conventional and traditional building materials. The shortage of building materials can be fulfilled by changing the paradigms of building materials use and nature of technology to be adopted. The major strategic needs to be adopted while using building materials and related technologies can be: (a) use of locally available building materials such as mud, stone, timber, lime depending upon geo-climatological needs and (b) introduction of appropriate and innovative technologies for utilising basic building materials like bricks, steel and cement in an effective, efficient and economic manner produced from agricultural and industrial wastes.

At present over 300 million tonnes of agro-industrials wastes are being produced per annum of various agricultural and process industries. Use of industrial wastes and by products is of paramount significance in developing building material components, providing a solution for the problems being faced in the housing industries. Presently, only about 20 per cent of the solid waste is being used out of the total waste generation in India while developed countries are utilising more than 45 per cent of the waste generated in their country.

The National Housing Policy adopted in both houses of Parliament in August,1994, also lays enormous stress on the use of innovative, cost effective yet sustainable building materials and technologies, which are environmentally appropriate, energy sensitive and ecological sustainable.

Building Materials and Technology

Simulating the increased production and availability of conventional materials like cement, steel and bricks as part of the industrial policy and promotion of small scale industries, even as steps are taken to economise the use of scarce building materials and to promote low cost and environmentally appropriate technology and use of indigenous resources.

Assured access of rural households to traditional materials including biomass and checks on indiscriminate commercial exploitation of forest resources, with due regard to envirofimental preservation.

Encouragement of public and private construction agencies

of materials based on industrial and agricultural wastes, these based on local resources and alternate appropriate technologies.

An alternative/proper building material based on technologies to ensure sustainable development suitable for developing economy is essential. The major parameters are given below:

i. Energy efficient
ii. Material production with out CO_2 emission
iii. Housing and building with out any adverse impact on environment
iv. Strong and durable building materials v. Resistant to natural disaster like earthquakes, cyclones, fire, etc.
vi. Affordable cost catering to the needs of masses vii Recycling of industrial wastes to obviate environmental threat.

Table: Availability of Agro-Industrial Wastes

Sl. No.	*Wastes*	*Source Million Tonnes/Annum*	*Availability*
1.	Fly Ash	Thermal Power Plants	120.0
2.	Blast Furnace Slag	Iron & Steel Industry	11.0
3.	Lime Sludge	Paper, Sugar, Fertilizer, Acetylene	5.0
Chromium Industry, Tanneries, Soda Ash			
4.	Phosphogypsum	Phosphoric Acid Fertilizer	6.0
5.	Fluorogypsum	Hydrofluoric Acid Industry	0.3
6.	Red Mud	Aluminium Industries	2.5
7.	Ferro Alloy and	Ferro Chrome Ferro-Manganese	3.5
Other Metallurgical Slags and Ferrosilicone Alloys and Smelting of Zinc and Copper			
8.	Cement Kiln Dust	Cement Industry	5.5
9.	Mine Tailings	Zinc, Copper, Gold, Iron ore Beneficiation	20.0
10.	Cinder	Thermal Power Plants and Furnaces	2.0
11.	Water Works Silts	Water Works	>10.0

Contd...

Sl. No.	Wastes	Source	Availability Million Tonnes/Annum
12.	Lime Kiln Rejects	Lime Kilns	Large Quantity
13.	Coal Washery Rejects	Coal Washeries	Large Quantity
14.	Rice Husk	Rice Shellers	50.0
15.	Bagasse	Sugar industries	90.0
16.	Marble dust	Marble industry	6.0
17.	Banna leaves/stalk	Banana plants	0.20
18.	Cocunut husk	Coir fibre industry	1.60
19.	Groundnut shell	Groundnut oil mills	11.0
20.	Saw mill waste	Saw mill/wood	2.0
21.	Sisal fibres	Sisalplantation	0.25
22.	Cotton stalk	Cottonplantation	0.20

Therefore, it is needed to have an enabling environment for use of these waste materials for building production. R&D has clearly established the use of fly ash, phosphogypsum, lime sludge, red mud among the industrial wastes and rice husk, bagasse, bamboo, coconut pith, rubberised wood other agro wastes for the production of value-added building materials. These building materials are being produced and used not to that extent where it has to be used. These building materials are based on industrial wastes which are contaminated with several harmful impurities, incompatible particle size, transport and handling problems. There are many myths and facts about the use of solid industrial wastes and their use in making new and profitable building materials. Some of the industrial wastes are described to remove these myths to instill confidence among the users of these building materials.

Fly Ash

Fly ash is an industrial waste of thermal power plants using pulverised coal as fuel. It is available to the tune of over 120 million tonnes per annum in India. Extensive work has been done at CBRI to find out various applications of fly ash. The quality of fly ash is judged by its lime reactivity. The lime reactivity of Indian fly ashes lie in between 2.5-7.5 N/mm^2.The major minerals present in the Indian fly ashes are quartz, calcite, haematite, magnetite and a small quantity of glass (20-30 per cent). The fly ash has been found suitable for making Portland- pozzolana cement (IS: 1489-1976, specification for Portland pozzolana cement), ready-mixed fly ash concrete, precast fly ash concrete building units (building blocks, flooring and roofing units), sintered lightweight aggregate, lime-fly ash cellular concrete, portland cement clinker, oil well cement, masonry cement, clay-fly ash and fly ash-lime bricks, high performance concrete, roller compacted concrete, in cement concrete roads, etc.

To enable production of building materials using fly ash, it is essential that fly ash should comply with the requirements of IS:3812-Part 1-2003 (specification for pulverised fuel ash, Part-1 Pozzolana and admixture in cement mortar and Concrete).

Nature of Fly Ash

Fly ash is refractory in nature and abrasive in nature and its particle size is very fine. The ultimate composition of fly ash is dependent on the original composition of minerals present in feed coals, source of supply of coals, degree of pulverisation, type of combustion equipment and the firing conditions. As compared to fly ashes available abroad, the Indian fly ashes contain relatively higher amount of silica, Al_2O_3 and carbon fuel and lower amount of CaO and SO_3. As in case of foreign fly ashes, the Indian fly ashes have been found to contain mullite, magnetitie, haematite, quartz and glass. Besides this, fly ashes may contain trace of some highly toxic metals like molybdenum, mercury, selenium, cadmium, their usual concentration may be in the following ranges.

Morphologically the fly ash particles may be irregularly shaped particles, solid spheres, plerospheres and cenosperes. The plerospheres contain many small spheres and cenospheres that are hollow in nature. The grinding of plerospheres and cenosperes may affect the physical properties of the fly ashes, i.e. the specific gravity, fineness water rquirement and pozzolanic reactivity.

Table: Chemical Requirements

Sl.No. Characteristics	Requirements Siliceous pulverised Fuel Ash	Calcareous Pulverised Fuel ash
1. Silicon dioxide(SiO_2) +AL_2O_3 + Fe_2O_3 in per cent by mass, Min	70	50
2. Silicon dioxide(SiO_2) in per cent by mass, Min	35	25
3. Reactive silica in per cent by mass, Min	20	20
4. Magnesium oxide (MgO) in per cent by mass, Max	5.0	5.0
5. Available alkalies as sodium oxide(Na_2O) in per cent by mass, Max	1.5	1.5
6. Total sulphur as sulphur trioxide (SO_3) in per cent by mass, Max	3.0	3.0
Total chlorides in per cent by mass, Max	0.05	0.05
7. Loss on ignition in per cent by mass, Max	5.0	5.0

Table: Physical Requirements

Sl.No. Characteristics	Requirements
i. Fineness-specific surface in m^2/kg by Blain's permeability method, Min	320
ii. Particles retained on 45 micron sieve (wet sieving)	34
iii. Lime reactivity- Av. Compressive strength, Min in N/mm^2	4.5
iv. Compressive strength at 28 days in N/mm^2, Min80 % of corresponding plain cement mortar cubes	
v. Soundness by autoclave test-Expansion in per cent, Max	0.8

Concentration Range	*Elements*
100 -1000 mg/kg of ash	B, Ba, Cu, Mn, Sr.
10 -100 mg/kg of ash	As, Cr, La, Mo, Ni, Pb, Th, U, Zn.
1 -10 mg/kg of ash	Cd, Ga, Sb, Se, Ti, V, Hg.

Fineness

Fineness of fly ash is a single important physical parameter which influences the activity of fly ash more than any other physical factor. It is commonly determined by air permeability method and is expressed as specific surface. The specific surface of Indian fly ashes range between 220-680 m^2/kg. Whereas the surface area of fly ashes produced in USA, France or Japan lie in between 200-607m^2/kg. The surface area of fly ashes of higher loss on ignition is also higher.

This is due higher level of unburnt carbon present in the fly ash. According to Brink and Halstead, the surface area determined by air permeability method is influenced by both external and internal surfaces of the particles of porous carbon. An increase in unburnt carbon content is accompanied by an increase in total surface area, even though the external surface area and the actual fineness decreases. As a result erroneous results are obtained. This could be the one of the reason which led to ASTM to drop the requirement on specific surface area in ASTM C 589-76.

Pozzolanacity

Degree of pozzolanic activity in fly ashes determines its suitability for use as pozzolana in the production of Portland-pozzolana cement or a part replacement of cement in mortars and concrete at the construction sites. The pozzolanic activity is generally increased with an increase in the fineness but no definite relationship was found between pozzolanic activity and either of the chemical and physical properties of Indian fly ashes. The pozzolanic activity of Indian fly ashes in general, comply with the requirement of 4.5 N/mm^2 specified in IS:3812 (Part-I)-2003.

Quality Variation in Indian Fly ashes

Despite lot of investigations on the utilisation of fly ashes, the actual utilisation of fly ashes in constructionwork has not made much head way. The main reasons for this state of affairs may be: (1) apprehensions regarding variability in the quality of fly ashes with consequent adverse effects on its performance, (2) absence of a marketing agency assuring to the supply of fly ashes conforming to the Indian Standard, and (3) handling and transportation of fly ash.

Control Measures

1. The carbon content of fly ash should not be more than specified value of 6.0 per cent.
2. The fineness requirement by wet sieving through 45 micron sieve and the measurement of unburnt carbon content may be considered as a control test for the fly ash. The lime reactivity may also be measured time to time.
3. The fly ash collected at the thermal power plants by electrostatic precipitators and mechanical collectors should not be mixed with the bottom ash. These fly ashes should be kept separate.
4. The burning of coal in the boilers should be efficient so that formation of more unburnt carbon in the fly ash may be avoided.

Phosphogypsum

Large quantity of by-product phosphogypsum is produced from phosphoric acid industry annually in India as a filter cake containing 15-30 per cent free moisture. The main impurities in the phosphogypsum are P_2O_5F, undecomposed phosphate rock, organic matter, quartz and residual acid.

In phosphogypsum, P_2O_5 generally exists as H_3PO_4, $Ca(H_2PO_4)_2$ H_2O, $CaHPO_4$. $2H_2O$ and $Ca_3(PO_4)_2$. P_2O_5 found adhered to the surface of gypsum crystals is water leachable, whereas P_2O_5 present in the crystal lattice is difficult to be removed by simple water washing. The fluorides are normally present as a mixture of NaF, Na_2SiF_6, Na_3AlF_6, Na_3FeF_6 and CaF_2 compounds which are derived from the rock phosphate on acidification with sulphuric acid. The organic matter is composed of aliphatic compounds of carbonic acids, amines and ketones.

These impurities adversely affect the setting and strength development of plaster and plaster products. For mass utilisation of by-product gypsum, it is imperative to remove or inactivate these impurities as far as possible. Enormous work has been accomplished at CBRI and elsewhere to beneficiate phosphogypsum for use in the manufacture of plaster and plaster products such as plaster boards, blocks, acoustic tiles, water-resistant binder, etc. The beneficiated techniques evolved at the Central Building Research Institute (CBRI), Roorkee, varied from simple washing to chemical leaching of phosphogypsum.

Every process has its own merit. The chemistry and technological aspects of these processes have been thoroughly investigated. Based on the basic work by wet sieve analysis technique, a pilot plant of capacity three tonnes per day for the beneficiation of phosphogypsum has been designed and developed in CBRI. The plant incorporates major equipment such as mixers, pumps, vibratory screen, centrifuge and rotary drier.

The waste phosphogypsum is beneficiated by passing the mixed gypsum-water slurry (1:3 by volume) through 300 micron sieve fitted in the vibratory screen whereby coarser fraction below 10.0 per cent richer in impurities retained over the sieve is discarded. The finer fraction is centrifuged and dried in the rotary drier at 110-120°C to get dried beneficiated phosphogypsum as per IS: 12679-1989, specification for by-product gypsum for use in plaster, blocks and boards. The plant is first indigenous development in the country. The plant can be scaled up to a capacity of 20 to 25 tonnes per day.

Table: Composition of Unbeneficiated/Beneficiated Phosphogypsum

Constituents,%	*Unbeneficiated Phosplwgypsum*	*Beneficiated*	*IS:12679- 1989 Limits*
P_2O_2	0.50-1.10	0.35-0.38	Max.0.40
F	0.55-1.12	0.35-0.39	Max. 0.40
Organic matter	0.22-0.59	0.05-0.09	Max.0.15
Na2O+K_2O	0.25-0.40	0.14-0.16	Max.30
pH	2.0-3.5	5.5-6.4	Min.5.0
Radioactivity	8-13 pCi/g	N.S.	

Production of β—Hemihydrate/Anhydrite Plasters

The industrial importance of the gypsum industry lies in its dehydration and the rehydration processes to give dihydrate gypsum of different properties. The samples of phosphogypsum and H-acid gypsum were calcined at 150°-160° C in the Gypsum Calcinator to form b-hemihydrate plaster or gypsum plaster. The plaster was ground to pass 150 micron IS sieve and tested for various physical properties as per is 2542-1984 (Part-I), specification for methods of test for concrete.

Fibrous Gypsum Plaster Boards

The gypsum plaster boards are lightweight boards generally made by reinforcing the teased sisal or sisal plus coir fibres in between two layers of the gypsum plaster slurry at 70-80 per cent water to give fluid workability. The plaster boards set in half an hour and are demoulded after one hour and then dried at 42±2°C either in the oven or in the sun.

The boards are known for their excellent fire resistant and acoustic properties and for their use in false ceiling, lightweight partitions and cladding. The influence of volume of fibres, length, density of matrix, impact strength, etc are reported. boards. The plaster boards have lower thermal conductivity than the conventional asbestos cement sheets generally used as false ceiling, thereby, facilitating better insulation of the gypsum boards. The boards are known for their excellent fire resistant and acoustic properties and for their use in false ceiling, lightweight partitions and cladding.

Gypsum Building Blocks

Gypsum blocks are lightweight, fire resistant building components suitable for use as lightweight partitions for non-load bearing walls. Gypsum blocks may be produced by mixing gypsum plaster with 70-80 per cent water to form thin slurry which is quickly filled in the moulds (400 mm x 200 mm x 100 mm). The moulds are opened after setting of blocks, dried and made ready for use. The gypsum blocks possess bulk density in the range 1,000-1,100 Kg/m^3 and the compressive strength 2.5 - 3.0 N/mm^2. The compressive strength conformed to the requirements laid down in IS: 2849 -1983, Specification for non-load bearing gypsum partition blocks (Solid and hollow type). The values of the bulk density and water absorption have not been specified.

Water-resistant Gypsum Binder

A water-resistant binder for external and internal applications has been developed by blending the ground gypsum plaster with fly ash or granulated blast furnace slag or red mud, OPC and a small quantity of a chemical retarder.

The gypsum binder has great potential in making not only masonry mortars and plastering but may be used in producing high strength fibre reinforced boards having low water absorption, in making gypsum blocks and water resistant tiles. There is a great demand for these materials for various applications in construction sector. The gypsum boards if produced using glass reinforced gypsum binder then it can be looked as an alternative to scarce wood. In the production of gypsum binder, lot of phosphogypsum waste can be used which can conserve high purity ($CaSO_4.2H_2O$ content > 90.0 per cent) natural gypsum, thus making available the high purity natural gypsum for other important industries such as dental, surgical and orthopedic plasters.

Table Properties of Gypsum Plaster Produced by Calacining Waste Gypsum

Sample desgn.	Consistency (%)	Setting time (Minutes) UR	Setting time (Minutes) R	Bulk density (Kg/m³) UR	Bulk density (Kg/m³) R	Compressive strength (MPa) UR	Compressive strength (MPa) R
Phosphogypsum							
Unprocessed	70.0	5.0	15.0	122	1210	9.80	8.90
Beneficiated	62.0	11.0	30.0	1250	1225	10.60	9.60
H-acid gypsum							
Unprocessed	65.2	6.0	28.0	1280	1200	8.20	8.00
Beneficiated	61.0	10.0	35.0	1250	1260	9.60	8.44
IS: 8272-1984 Limits	-	20-35		-		Min.7.6	

Table Properties of Fibrous Gypsum Plaster Boards Made from Phosphogypsum Plaster

Characteristics	*Result*	*Asbestos-cement*	*IS: 2095 (Part3)-1996*
Transverse strength, N	450-550	500-600	Min.340
Bulk density, Kg/m³	1000-1100	1600-1880	
Deflection, mm	1.5	—	Max. 19.0
Hardness, mm	Passes		Max. 5.0.
Thermal conductivity, Kcal/m/hr/°C	0.12-0.18	0.24-0.3	

Another application in which gypsum binder can be used to a great extent is the mining where the binder has immense potential as mine stopping material, hence, a tremendous saving in cost as well as conservation of natural resource may be realised.

The durability of the binder as studied by alternative wetting and drying and heating and cooling cycles and by the long term performance in water showed much better stability against the water than the plain plaster. The gypsum binder is eminently suitable for masonry mortars (mixes 1:4, 1:5 binder-sand- Fineness modulus (1.22, 2.2) owing to high workability, high strength values and low cost than the 1:6 cement-sand mortars. These mortars fall in the grade of 2.0 to 3.0 and 3.0 to 5.0 N/mm^2. Glass reinforced gypsum binder composites of density 1.58 g/cc, flexural strength 18-20 MPa, tensile strength 16-18 N/mm^2, impact strength 14-16 Nmm/m^2 and thermal conductivity 0.9 Kcal/ m./ h./°C have been developed from the gypsum binder.

The engineering properties of the binder composites suggest their possible use as an alternative to timber in door shutters, structural partitions, ceilings, etc.

Table: Physical Properties of Water-Resistant Gypsum Binders

Properties Gypsum Binder with fly ash	*Gypsum Binder with slag and*	*Binders Plain Gypsum plaster red mud*	
1. Fineness, m^2/kg (Blaine)	318	310	300
2. Setting time, Min. Initial Final	95	50 150	25 100
3. Compressive strength, MPa, (28 days)	22.0	34.2	13.3
4. Soundness, mm (cold expansion)	0.88	1.0	—
5. Water absorption, %	< 10.0	!4.0	Leaching

Multiphase Plaster

The normal plaster on setting and hardening produce low density and medium strength. In contrary to it, anhydrite plaster sets slowly and attains maximum strength at 28 days of curing. Multiphase plaster has quick initial setting and gradual final setting (25-30 minute) and these plasters develop quite high strength (20-25 MPa) which is not possible with ordinary plaster of Paris. Most of the work reported on multiphase plaster is patented and little information is available. Moreover, the multiphase plaster have been produced using natural gypsum. It was decided to produce multiphase plaster which does not require retardation or activation and gives properties required for plastering. The multiphase plaster has been produced from the phosphogypsum heated at 150 -1000°C wherein a jumble of â-hemihydrite, soluble and insoluble anhydrite plasters are formed. The MPP was also produced using fluorogypsum as the anhydrite component. The MPP was tested for properties like consistency, setting time, bulk density, compressive strength and chemically combined water as per relevant standards and test procedures. Data show an increase in compressive strength, bulk density and chemically combined water with increase in curing period. Addition of small quantity of activator (mix. C) does not enhance the strength appreciably. The attainment of strength is remarkably higher than the plaster of Paris (POP). It can be seen that the strength development in the MPP based on fluorogypsum is slightly higher than the other waste gypsum based plasters.

Table: Properties of multiphase plaster produced from phosphogypsum

Mix desgn.	*Properties*											
	Consis-tency (%)	*Setting time (Minutes)*		*Compressive strength (MPa)*			*Bulk density (glcc)*			*Chemically Combined water (%)*		
		Intial	*Final*	*3d*	*7d*	*28d*	*3d*	*7d*	*28d*	*3d*	*7d*	*28d*
A	46	15	20	5.6	6.8	15.6	1.0	1.1	1.1	5.6	7.0	14.5
B	40	15	25	5.2	6.0	16.8	1.2	1.1	1.2	6.9	8.8	16.9
C	38	20	28	6.0	6.9	17.8	1.1	1.2	1.2	7.1	8.6	17.9
POP	56	8.0	-	10.8	(24hrs.)		0.9	1.0	1.1	4.5	18.2	18.8

The multiphase plaster can be used in plastering both for finish coat (3mm) as well as base coat (9mm-MPP:Sand) in buildings. The highly glossy surface of MPP may replace "Neeru" (a lime based material) used over coarse plastering in cosmopolitan cities. Multiphase plaster may be blended with the lightweight exfoliated vermiculite, perlite or any compatible aggregate for making lightweight thermal insulative plasters. Amix of IPP-Expanded Perlite Aggregate -75:25 by wt. was found effective to give density 1000-1100 kg/cm^3 and low thermal conductivity.

Flooring Tiles from Waste Gypsum: The flooring tiles of sizes 200mm × 200mm × 20mm and 300mm × 300mm × 20 mm were cast by vibration moulding of the mix containing phospho/fluoro anhydrite powder with different pigments, polymers, a small quantity of glass fibre and coloured stone chips at normal consistency. The fly ash and red mud industrial solid wastes were also added to the anhydrite mixes for moulding these tiles. These tiles after demoulding were cured in high humidity (over 90 per cent), dried at 42±2°C and tested for properties such as flexural strength, compressive strength, water absorption, wear resistance and porosity as per IS:1237-1980, specification for cement and concrete flooring tiles.

Table: Properties of flooring tiles produced from phosphogypsum / fluorogypsum

Property Value	*IS:1237-1980*	*Limits*
Perpendicularity,%	0.7	Max. 2.0
Straightness, %	0.6	Max. 1.0
Flatness, mm	0.4	Max. 1.0
Flexural strength,N/mm^2	11-15	Min. 3.0
Compressive strength, N/mm^2	30-40	N,S.
Water absorption, %	2.0-3.5	Max. 10.0
Wear resistance, mm	2.0-4.0	Av.3.5, Individual 4mm, Gen. purpose
Porosity	5.0-8.0	N.S.

N.S.= Not Specified.

As Filler: Researches have been reported for developing thermally modified phosphogypsum for use as filler for low density polythene polypropylene, polymide 6 and PVC compositions. Filling of polyolefins with phosphogypsum is intended to enhance polyolefines mechanical tensile properties. The phosphogypsum filled compositions has great scope in talc and automotive industries.

Blast Furnace Slag

In India about 18 million tonnes of molten blast furnace slag is produced every year and out of this 5.0 million tonnes is granulated slag. Granulated blast furnace slag is a glassy material having latent hydraulic propeties. Extensive work has been carried out at CBRI to make portland slag cement (PSC) from the slag available from the steel plants in India conforming to IS 455-1989, specification for Portland slag cement. PSC has also been produced from the high manganese slags (Mn_2O_3 up to 6.17 per cent) available from Rourkela steel plant. PSC can be used in all types of civil works. It is eminently suitable for marine and mass concrete construction due to high-sulphate-resistance and low heat of hydration besides wide use in water-retaining and water tight structures in construction of rivers, ports, harbours, hot springs and sewer systems of factories and mines.

The granulated blast furnace slag has been used for making super sulphated cement (SSC). It has been found that the slag can be activated with 20-25 per cent phospho/ fluoroanhydrite as against the commonly recommended value of 10-15 per cent. The SSC conformed to the requirements of IS:6909-1990, specification for super-sulphated cement. Some of the important advantages of SSC are its low cost with saving in energy, increased resistance to sulphate and lower heat of hydration compared to Portland cement. A new product namely 'Slagstar' has been reported from Austria which is based on ground granulated blast furnace slag gypsum and alkaline additives. Due to its sulphate activation, no burning process is required for the production compared to the production of Portland cement.

Recently a low heat sulphate resistant (LHSR) cement has been developed at CBRI wherein low temperature form of the calcined gypsum, i.e. b- hemihdrate plaster ($CaSO_4.0.5\ H_2O$) was used in place of anhydrite ($CaSO_4$) formed at 800-850 °C. Cement saves lot of energy. Moreover, the cement is sulphate resistant as it showed better sulphate resistance in sulphate solutions than the traditional portland cement. LHSR cements A and B contained 75.0 per cent granulated blast furnace slag, 15 per cent cement clinker and 10 per cent hemihydrate plaster. The LHSR cement C contained 15 per cent anhydrite in place of hemihydrate.

Table: Properties of LHSR cements

Cement			*Properties design.*			
Setting		*time*	*Compressive strength*			*Soundness*
(Minutes)			*(MPa)*			*(Coldexpn.)*
Initial		*Final*	*Time of hardening*			*(mm)*
3 day 7day 28 day						
A	10	65	16.1	23.7	35.4	1.0
B	55	106	16.9	38.4	56.1	1.2
C	110	17.5	28.0	46.0	1.6	IS: 6909 - 1990
Limits	Min.30	Max.600	Min.15	Min.22	Min.30	Max.5.0

Ground granulated blast furnace slag has been used to make pressed bricks by activating with hydrated lime. An optimised moist slag-lime mix was blended with 50,60,70 and 80 per cent of the sand (Fineness modulus 2.2) and pressed in mould at 5.0 -6.0 N/mm^2 pressure and cured. A mix of slag lime - sand, 30:70 (by mass) developed compressive strength 12.1N/mm^2, water absorption 10 per cent and bulk density 1940 Kg/m^3. The bricks are suitable for use in construction work.

Lime Sludges

Large quantities of lime sludges are produced as a by-product from the sugar, calcium carbide, acetylene, soda ash, tannery and paper industries as calcium carbonate. The main impurities in lime sludges are sugar, P_2O_5 and alkalies. The presence of impurities in lime sludges is considered to be responsible for the poor strength although they have the beneficial effect of lowering the clinkering temperature. These sludges have been used in different ways in the production of cement and cementitious materials.

Lime sludges have been used in the manufacture of masonry cement as a replacement to the lime powder. Based on the studies conducted at CBRI it has been found that masonry cement can be produced from lime sludges obtained from calcium carbonate, acetylene, sugar industries conforming to IS:3466-1988, specification for masonry cement.

The hydraulic binder having some of the properties similar to Portland cement has been developed from the rice husk and by-product lime of sugar, acetylene and paper. Rice husk acts as an integral fuel and provides *in situ* silica for the lime produced during the firing at 900-950°C. The material produced is a soft powder. The activity of hydraulic binder increases with the fineness.

The setting characteristics of the binder are within the time limits similar to those of hydraulic cements, IS:4031-1988, Methods for physical testing of hydraulic cements, with the improved water retention, workability, water tightness, etc. On the basis of above properties, The hydraulic binder has been found suitable for use in making masonry mortars, plasters, foundation concrete and concrete for sub-flooring and terracing works.

An rapid setting activated lime-pozzolana mixture (ALPM) based on the calcined lime sludges and fly ash/surkhi/ metakaolin conforming to IS:4098-1983 has been developed at CBRI. The lime-pozzolana mixtures possess low density (750-900 kg/m^3), high water-retentivity (70%) and have low expansion or shrinkage (0.02 to 0.03%). They can be converted into mortars and plasters directly through mixing with sand and water.

Investigations carried out at CBRI indicate that bagasse ash possesses good pozzolanic activity. It was found that it contains two-third silica most of which is in amorphous state. Bagasse ash, therefore, is a pozzolanic material like fly ash and other traditional pozzolanas. Studies have been undertaken to produce quick setting lime-bagasse ash mixture by inter grinding a mixture of hydrated lime, bagasse ash along with 10-15 per cent Portland cement and 4.0 per cent gypsum to a fineness of passing 90 per cent passing through *75* micron sieve. Such a mixture would have potential application as a mortar and plaster particularly in rural areas.

For the production of good quality lime from the lime sludges CBRI has designed lime kiln of production capacity 5, 10 and 15 tonnes per day. For dust free production of hydrated lime, CBRI has developed lime hydrator of capacity 20 TPD. Over two dozen such hydrators are in use in the country for several years.

Lime is a major building material and its production results in evolution of some pollutants from the kilns. These pollutant are obnoxious gases such as SO_2, CO, CO_2 organic volatile and coal dust emitting from the top layer of the lime kiln. The Central Pollution Control Board (CPCB) has prescribed a permissible value of the particulate matter concentration of the order 500 mg/nm^3.

The suspended particulate matter (SPM) and the tarry organic matter have been identified to the significant pollutants emitted from the lime kilns. The same were required to be brought within the tolerance level. A system incorporating a double-deck packed bed scrubber-cum-entrainment separator was developed at CBRI to control emissions of dust particles and hydrocarbon tarry matter from the lime kilns.

It consists of two fixed-bed chambers containing lime stone packing. The lower one serves as the counter current gas-liquid (water) scrubber. The upper bed works as demister vis-a-vis entrainment separator for the removal of moisture and tarry vapours from the up flowing gases. The gases after cleaning escape to the atmosphere through an induced draft fan installed at the bottom of the stack. The performance of the control device was evaluated.

The SPM (comprising kiln dust and organic volatile) level 1000-2000 mg/nm^3, was lowered down to less than the limiting value of 500 mg/nm^3 as prescribed by the CPCB, Delhi, with overall system efficiencies for dust removal being in the range of 80-85 per cent. The system is already in use in several lime kilns.

Red Mud

Bauxite red mud is a by-product released during the process of aluminium extraction from caustic leaching of bauxite and it poses serious environmental concern for storage and disposal. Studies carried out by Li showed that the major solid component of the red mud waste produced by the Bayer's process contains iron minerals geothite, hematite and magnetite. The red mud aluminium industry is an important factor in assessing the environmental impact of red mud disposal. A new construction materials produced by blending red mud waste with 2 of lime or Portland cement of improved strength and low leachability has been reported The high strength of binder allow their applications in road, airfield bases, foundations, etc.

Tailings from Metal Processing Industry

Zinc is an basic metal needed for various applications in metallurgical, chemical and textile industries. Geothite process (FeOOH) and Jarosite process ($MFe_3(SO_4)_2(OH)_6$ are used for the manufacture of zinc metal. Investigations have been carried out to utilise mining wastes generated from copper, zinc, gold, nickel and molybdenum ore beneficiation processes in the production of masonry cements, sand-lime bricks and cellular concrete at CBRI.

The zinc tailings obtained during the beneficiation of zinc, generally contain silica and dolomite. Masonry cement has been produced by grinding these tailings with Portland cement and sucrose as set regulator to a fineness of 500 $m^{2/}$ kg.

Table: Properties of Gold Tailings Based Masonry Cement

Properties Requirements	*Masonry cement*	*IS: 3466-1988*
Fineness (m^2/kg)	612	Min. 500
1. Setting time (Hours)		
Initial	2.5	Min. 1 Hour 30 Minutes
Final	5.0	Max. 24.0 Hours
3. Compressive strength		
(MPa) 7 days	3.6	Min. 2.5
28 days	6.8	Max. 5.0
4. Water retention	66	Min. 60
5. Soundness, Lechatelier's expansion ((mm)	1.1	Max. 10
6. Air content	5.8	Max.6.0

Apprehensions and Reality about Asbestos

Asbestos Cement Industry, besides catering to a wide range of domestic and commercial requirements of the society at large, has been under constant threat of several myths and negative propaganda. The awareness levels about asbestos are at very low level and border on ignorance that most people do not even know that there are different type of asbestos and risk potential differs by product and process.

Moreover, majority of people even do not know that asbestos cement products, as used in India today, do not have any risk potential to the users.

Asbestos cement products contain only 8 per cent of white asbestos called Chrysotile while the balance consists of fly ash (30%), cement (40%), water (20%) and pulp (2%). White asbestos is a naturally occurring mineral with no known harmful effect, under controlled uses in asbestos cement manufacturing. Asbestos is divided into two types according to mineralogical classification - Serpentines and Amphiboles.

Chrysotile comes from Serpentine while Amphiboles yield Anthrophylite, Crocidolite, Amosite, Actinolite and Tremolite. All the six types of asbestos are hydrated silicates but differ considerably in their metal content. There are marked difference in the degree of hazard among the different asbestos fibres types.

There are three basic asbestos-related diseases called asbestosis, lung cancer and Mesothelioma. Out of these, Mesothelioma is not reported when only Chrysotile is used. The blue asbestos is not used in India and the Government has banned the import of its fibre in 1994, following the recommendation of the industry.

Studies by Central Labour Institute, Mumbai (2000 and 2001) and Regional Labour Insitute, Chennai (1989) pointed out that not even a single case of lung cancer was noted in Chrysotile asbestos cement products manufacturing industry workers in India.

With regard to asbestosis, a number of studies in the US, South Aferica and Canada found out that the disease would occur with a prolonged exposure of 5 to 20 fibres per cc over a period of 40 years. Even the fibre dust count at different places in Indian asbestos cement plants is less than 1 fibre per cc. Even at construction site, where fixing of AC sheets is done using hand driven tools, dust concentration does not exceed 0.1 fibre per cc. The Bureau of Indian Standard IS:11451-1986, Recommendations for Safety and Health Requirement Relating to Occupational Exposure to Asbestos, has specified this limit as 2 fibres/ml apparently for all types of asbestos.

The Environmental Protection Act-1986 (Schedule VI, Part D), has given 4 fibres / ml limit for general emission of asbestos. According to Datta, with the controlled use of asbestos, the environment, environmental pollution and health risk is extremely low.

Indian asbestos cement industry's annual turnover is estimated to be in the range of Rs.1000-1100 Crore. Its offering to the society are manifold in employment, environment protection and low consumption of energy. The industry has been providing a direct employment to over 30,000 people and indirect employment to over one lakh. It has been fulfilling the needs of Indian society, in construction, utilities and sanitation sectors.

Radioactivity in Building Materials

Generally people spend their time indoors and natural rdioactivity in building materials is a source of indoor radiation exposure. Indoors elevated dose rates arise from high activities of radionuclides in building materials. Most building materials contain naturally occurring radioactive elements the most important of which are 40K and their isotopes ^{232}Th and^{226} Ra. The presence of these radioisotopes in the building materials causes external exposure to the inert gases of ^{222}Rn ^{220}Rn and their daughters in the house.

Chronic exposure of human beings to low doses of ioning radiation can cause health damages which may appear 5-30 years after the exposure and sometimes malignant diseases may be contracted by the people. To minimise the exposure of population to the radiation, there is need to control and to limit the content of radioactive materials in dwellings. There is a growing tendency to use new recycled materials such as fly ash and phosphogypsum with presence of radioactivity.

Radioactivity in building materials became focus of regulatory interest with the publication of International Atomic Energy Agency SS 115 and subsequent publication of European Council Directive 96/29/EUROTOM.

The importance of the problem can be gauged from the fact that some of the countries like Poland, Italy and South Africa have formed legal regulations or standards in terms of radioactivity for raw materials being used as building construction materials, phosphate and fertilizer industries. Radionulclide concentration in phosphogypsum are variable and depend on the radionuclide content of the original ore and the method of production.

Indian Scenario

In India, phosphogypsum is used to make construction material such as gypsum plaster, gypsum ceilling tiles, boards, panels and marbles/blocks.

It is promoted extensively by the building materials and Technology Promotion Council (BMTPC), Ministry of Urban Development. Based on BMTPC's recommendations, the government has even exempted phosphogypsum from excise duty.

Hence, in addition to what is produced in the country, phosphogypsum is also imported from countries such as Jordan and Morocco. It is interesting to note that the phosphate rock imported from Morocco contains radium of the order of 46 picocurie per gram (pCi/g)-according to United Nations Scientific Commission for Emissions, while that of India has only 4pCi/g. As the imported phosphate rock contains higher level of radioactivity, the radon level is also high exposing people to higher level of radiation.

The Environmental Protection Agency (EPA) has put up a safe limit of 10 pCi/g where as International limit prescribed by Euratom (European Atomic Commission) has fixed the upper limit of 13.5 pCi/g.

Till date there is no unanimity on the "Safe Limit" for radioactive exposure due to PG. Dr. B.V.Swaminathan, Chief, Fertilizer Association of India (FAI) New Delhi, "When the International community can not come to a conclusions, how can India expect to do so? While the BMTPC, refused to acknowledge that Indian PG is radioactive.

The issue of rock phosphate imports is assigned to the Ministry of Environment and Forests (MEF) and Ministry of Chemicals and Fertilizers (MCF).

Table: Radioactivity in Common Industrial Wastes used in Building Material

Common building materials and Industrial By-products	*Typical Activity Concentration*				*Maximum Activity Concentration*			
	^{226}Ra	232 Th	^{40}K	Radium ^{226}Ra Equi.	^{232}Th		^{40}K	Radium Equi
Concrete	40	30	400	114	240	190	1600	635
Aerated and light wt. Concrete	60	40	430	150	2600	190	1600	2995
Clay red bricks	50	50	670	173	200	200	2000	640
Sand-lime bricks	10	10	330	50	25	30	700	122
Natural building stones	60	60	640	195	500	310	4000	1251
Natural gypsum	10	10	80	30	70	100	200	228
Phosphogypsum plaster board	4-700	1-53	—	—	—	—	—	—
Industrial By-Products								
Phosphogypsum	390	20	60	423	1100	160	300	1352
Blast furnace Slag	270	70	240	373	1100	300	1500	1644

Conclusions

From the data presented above, the following valuable conclusions can be drawn:

1. There is immense shortage of conventional building materials like bricks and timber with many environmental and ecological intricacies.
2. Several industrial wastes like fly ash, slag, phosphogypsum, red mud, lime wastes, etc. are available with different characteristics which need immediate effective disposal to upgrade environment.
3. There are apprehensions in using wastes in view of certain physical parameters like particle size, impurities radioactivity present in them.
4. Numerous myths of using fly ash and factual situation has been discussed to instill faith amongst the users.
5. Doubts and remedies about use of asbestos in India and the radioactivity in building materials, in general and phosphogypsum, in particular has been discussed to motivate people to use building materials produced from industrial wastes for cleaner environment.

9

Sewage Treatment

Bacteria in Waste Treatment

Bacteria are very important in sewage treatment. Standard sewage treatment involves multiple processes. It usually starts with settling during which large items sink to the bottom. Next, air is bubbled through the sewage. This so-called aerobic phase encourages oxygen-using bacteria to break down organic material in the sewage, such as human wastes, to acids and CO_2. Most disease-causing organisms are also killed in this phase. The sewage sludge left behind is attacked in a subsequent phase by anaerobic bacteria (bacteria that cannot tolerate oxygen). These bacteria break down the sludge to produce methane gas, which can then be used as a fuel to power the treatment facility. In treatment plants today, this anaerobic phase sometimes precedes the aerobic phase.

Bacteria are also effective in cleaning up harmful wastes through bioremediation. In this process, bacteria and other microorganisms convert toxic or otherwise objectionable wastes, such as pesticides and oil spills, to harmless or even useful products.

Bacteria have been at the centre of recent advances in biotechnology—the creation of products for human benefit through the manipulation of biological organisms. Biotechnology itself dates back at least as far as ancient Egyptian civilization. Paintings on the walls of Egyptian tombs depict the brewing of beer, which uses microorganisms in the fermentation process. However, the existence of bacteria did not become known until the development of sufficiently powerful microscopes in the late 1600s. During the centuries that followed, scientists became aware that living organisms were responsible for many biotechnological processes.

Biotechnology grew steadily during the 20th century. In the 1970s, scientists used information about replication of viruses and bacteria and about DNA synthesis (manufacture) to begin the genetic engineering of bacterial cells. When scientists combined human DNA with the DNA in bacterial cells, recombinant DNA technology was born. Human DNA is the "recombinant." DNA contains the instructions for creating proteins. With their recombinant DNA, bacteria became factories for turning out human proteins, such as the hormone insulin or antibodies that fight disease. Because they multiply so rapidly, bacteria produce multiple copies of proteins in a short time. The process of taking genetic information from one organism and placing it in a different organism was patented by American biochemists Stanley Cohen and Herbert Boyer in 1980. The genetic revolution was underway.

Due to the increase in population, most of the sewage treatment plants get challenged with hydraulic and shock loads. The pollutants in domestic wastewater not only arise from domestic and food service functions, but also from some home industries and small-scale industrial units. It is practically impossible to control the varied nature of the wastewater entering the treatment plants. The operators find it very difficult to run a stable plant while discharging water conforming to tight norms.

Currently a lot of energy is spent in providing the dissolved oxygen to the generic microbes, without getting the desired results. Though it is known that the biodegradation is done by the biomass, very little attention is given to control the type of bacteria that dominates the treatment system. It is not only how many bacteria but also the type and functionality of the bacteria that is very important in a sewage treatment plant. Several times chemical treatment approaches have been taken for remediation of sewage treatment plants and for clearing collection lines but they all have disadvantages.

Biotechnology has always played a key role in removing organic solids like human waste from the millions of litres of waste water generated every day in Australia. Other contaminants, though, like phosphorus and nitrogen, have often been discharged into rivers and other waterbodies where they can disrupt the delicate ecological balance.

Being nutrients, phosphorus and nitrogen can cause excessive plant and algal growth. Overgrowth of aquatic plants can choke rivers and dams, and algae can produce toxins that poison fish and livestock. To remove nutrients from waste water, sewage farms have been using costly chemicals. Biotechnology, however, offers a cheaper and cleaner alternative. An extra anaerobic stage (one without oxygen) is added to the beginning of the sewage treatment process to help break down the organic materials. In the next aerobic phase, where oxygen is available, bacteria responsible for phosphorus removal can proliferate and consume the readily available organic food source.

During the sewage treatment process, bacteria also help to convert nitrogenous compounds into gaseous forms of nitrogen which are allowed to escape into the atmosphere.

Before the word biotechnology was invented, a sewage plant required bacteria in order to function properly. Since then, engineers have developed more sophisticated methods and the use of these invisible but indispensable aids has been improved by both scientists and construction engineers. Membrane filtration systems such as those used in the plants of Wehrle Umwelt GmbH in Emmendingen for the purification of highly loaded industrial wastewater, a process in which the minute but important bacteria do not get lost, are of increasing importance in wastewater treatment.

Sewage treatment, or domestic wastewater treatment, is the process of removing contaminants from wastewater, both run-off and domestic. It includes physical, chemical and biological processes to remove physical, chemical and biological contaminants. Its objective is to produce a waste stream (or treated effluent) and a solid waste or sludge also suitable for discharge or reuse back into the environment. This material is often inadvertently contaminated with toxic organic and inorganic compounds.

Sewage is created by residences, institutions, and commercial and industrial establishments. It can be treated close to where it is created (in septic tanks or on site package plants and other aerobic treatment systems), or collected and transported via a network of pipes and pump stations to a municipal treatment plant. Sewage collection and treatment is typically subject to local, state and federal regulations and standards (regulation and controls). Industrial sources of wastewater often require specialised treatment processes.

Typically, sewage treatment involves three stages, called *primary*, *secondary* and *tertiary treatment*. First, the solids are separated from the wastewater stream. Then dissolved biological matter is progressively converted into a solid mass by using indigenous, water-borne bacteria. Finally, the biological solids are neutralised then disposed of or reused, and the treated water may be disinfected chemically or physically (for example by lagoons and microfiltration). The final effluent can be discharged into a stream, river, bay, lagoon or wetland, or it can be used for the irrigation of a golf course, green way or park. If it is sufficiently clean, it can also be used for groundwater recharge.

Raw influent (sewage) is the liquid waste from toilets, baths, showers, kitchens, sinks, etc. Household waste that is disposed of via sewers. In many areas sewage also includes some liquid waste from industry and commerce.. In the UK, the waste from toilets is termed foul waste, the waste from items such as basins, baths, kitchens is termed sullage water, and the industrial and commercial waste is termed trade waste.

The division of household water drains into greywater and blackwater is becoming more common in the developed world, with greywater being permitted to be used for watering plants or recycled for flushing toilets. A lot of sewage also includes some surface water from roofs or hard-standing areas. Municipal wastewater, therefore, includes residential, commercial, and industrial liquid waste discharges, and may include stormwater run-off. Sewage systems capable of handling stormwater are known as combined systems. Such systems are usually avoided since they complicate and thereby reduce the efficiency of sewage treatment plants owing to their seasonality. The variability in flows also leads to often larger than necessary, and subsequently more expensive, treatment facilities. In addition, heavy storms that contribute more flows than the treatment plant can handle may overwhelm the sewage treatment system, causing a spill or overflow (called a combined sewer overflow, or CSO, in the US). It is preferable to have a separate storm drain system for stormwater in areas that are developed with sewer systems.

The construction of combined sewers is a less common practice in the United States and Canada than in the past and is no longer accepted within building regulations in the UK and other European countries. Instead, liquid waste and stormwater are collected and conveyed in separate sewer systems, referred to as sanitary sewers and storm sewers in the US and as foul sewers and surface water sewers in the UK.

As rainfall runs over the surface of roofs and the ground, it may pick up various contaminants including soil particles, (sediment), heavy metals, organic compounds, animal waste, and oil and grease. Some jurisdictions require stormwater to receive some level of treatment before being discharged directly into waterways. Examples of treatment processes used for stormwater include sedimentation basins, wetlands, buried concrete vaults with various kinds of filters, and vortex separators (to remove coarse solids).

The site where the raw wastewater is processed before it is discharged back to the environment is called a wastewater treatment plant (WWTP). The order and types of mechanical, chemical and biological systems that comprise the wastewater treatment plant are typically the same for most developed countries:

- Mechanical treatment;
  — Influx (Influent)
  — Removal of large objects
  — Removal of sand and grit
  — Pre-precipitation
- Biological treatment;
  — Oxidation bed (oxidising bed) or aeration system
  — Post-precipitation
  — Effluent
- Chemical treatment (this step is usually combined with settling and other processes to remove solids, such as filtration. The combination is referred to in the US as physical-chemical treatment.).

Treatment Stages

Primary Treatment

Primary treatment removes the materials that can be easily collected from the raw wastewater and disposed of. The typical materials that are removed during primary treatment include to fats, oils, and greases (also referred to as FOG), sand, gravels and rocks (also referred to as grit), larger settleable solids including human waste, and floatable materials. This step is done entirely with machinery, hence the name mechanical treatment.

Influx (influent) and Removal of Large Objects: In the mechanical treatment, the influx (influent) of sewage water is strained to remove all large objects that are deposited in the sewer system, such as rags, sticks, condoms, sanitary towels (sanitary napkins) or tampons, cans, fruit, etc. This is most commonly done with a manual or automated mechanically raked screen. This type of waste is removed because it can damage or clog the equipment in the sewage treatment plant.

Sand and Grit Removal: Primary treatment typically includes a sand or grit channel or chamber where the velocity of the incoming wastewater is carefully controlled to allow sand grit and stones to settle, while keeping the majority of the suspended organic material in the water column. This equipment is called a detritor or sand catcher. Sand grit and stones need to be removed early in the process to avoid damage to pumps and other equipment in the remaining treatment stages. Sometimes there is a sand washer (grit classifier) followed by a conveyor that transports the sand to a container for disposal. The contents from the sand catcher may be fed into the incinerator in a sludge processing plant, but in many cases, the sand and grit is sent to a land-fill.

Screening and Maceration (Raw Sewage Pumping): The grit free liquid is then passed through fixed or rotating screens to remove floating and larger material such as rags and smaller particulates such as undigested food particles or materials that were not adequately pulverised with residential food disposal units. Screenings are collected and may be returned to the sludge treatment plant or may be disposed of off site by landfilling or incineration. Maceration, in which solids are cut into small particles through the use of rotating knife edges mounted on a revolving cylinder, is often used in plants that do not remove the larger particulates at this stage of treatment and that can process this particulate waste.

Sedimentation: Many plants have a sedimentation stage where the sewage is allowed to pass slowly through large tanks, commonly called "primary clarifiers" or "primary sedimentation tanks". The tanks are large enough that faecal solids can settle and floating material such as grease and oils can rise to the surface and be skimmed off. The main purpose of the primary stage is to produce a generally homogeneous liquid capable of being treated biologically and a sludge that can be separately treated or processed. Primary settlement tanks are usually equipped with mechanically driven scrapers that continually drive the collected sludge towards a hopper in the base of the tank from where it can be pumped to further sludge treatment stages.

Secondary Treatment: Secondary treatment is designed to substantially degrade the biological content of the sewage such as are derived from human waste, food waste, soaps and detergent. The majority of municipal and industrial plants treat the settled sewage liquor using aerobic biological processes. For this to be effective, the biota require both oxygen and a substrate on which to live. There are number of ways in which this is done. In all these methods, the bacteria and protozoa consume biodegradable soluble organic contaminants (e.g. sugars, fats, organic short-chain carbon molecules, etc.) and bind much of the less soluble fractions into floc. Secondary treatment systems are classified as fixed film or suspended growth. Fixed-film treatment process including trickling filters and rotating biological contactors where the biomass grows on media and the sewage passes over its surface. In suspended growth systems— such as activated sludge—the biomass is well mixed with the sewage and can be operated in a smaller space than fixed-film systems that treat the same amount of water. However, fixed-film systems are more able to cope with drastic changes in the amount of biological material and can provide higher removal rates for organic material and suspended solids than suspended growth systems.

Roughing filters are intended to treat particularly strong or variable organic loads, typically industrial, to allow them to then be treated by conventional secondary treatment processes. Characteristics include typically tall, circular filters filled with open synthetic filter media to which wastewater is applied at a relatively high rate. Designed to allow high hydraulic loading and a high flow-through of air. On larger installations, air is forced through the media using blowers. The resultant wastewater is usually within the normal range for conventional treatment processes.

Activated Sludge: Activated sludge plants use a variety of mechanisms and processes to use dissolved oxygen to promote the growth of biological floc that substantially removes organic material. It also traps particulate material and can, under ideal conditions, convert ammonia to nitrite and nitrate ultimately to nitrogen gas.

The carbon adsorption following biological treatment was particularly effective in reducing both the bod and cod to low levels. A fluidised bed reactor is a combination of the most common stirred tank packed bed, continuous flow reactors. It is very important to chemical engineering because of its excellent heat and mass transfer characteristics. In a fluidised bed reactor, the substrate is passed upward through the immobilised enzyme bed at a high velocity to lift the particles. However, the velocity must not be so high that the enzymes are swept away from the reactor entirely. This causes high mixing, this type of reactors are highly suitable for the exothermic reactions. It is most often applied in immobilised enzyme catalysis.

Filter Beds (Oxidising Beds): In older plants and plants receiving more variable loads, trickling filter beds are used where the settled sewage liquor is spread onto the surface and of a deep bed made up of coke (carbonised coal), limestone chips or specially fabricated plastic media. Such media must have high surface areas to support the biofilms that form. The liquor is distributed through perforated rotating arms radiating from a central pivot. The distributed liquor trickles through this bed and is collected in drains at the base. These drains also provide a source of air which percolates up through the bed, keeping it aerobic. Biological films of bacteria, protozoa and fungi form on the media's surfaces and eat or otherwise reduce the organic content. This biofilm is grazed by insect larvae and worms which help maintain an optimal thickness. Overloading of beds increases the thickness of the film leading to clogging of the filter media and ponding on the surface.

Biological Aerated Filters: Biological Aerated (or Anoxic) Filter (BAF) combines filtration with biological carbon reduction, nitrification or denitrification. BAF usually includes a reactor filled with a filter media. The media is either in suspension or supported by a gravel layer at the foot of the filter. The dual purpose of this media is to support highly active biomass that is attached to it and to filter suspended solids. Carbon reduction and ammonia conversion occurs in aerobic mode and sometime achieved in a single reactor while nitrate conversion occurs in anoxic mode. BAF is operated either in upflow or downflow configuration depending on design specified by manufacturer.

Membrane Biological Reactors: Membrane biological reactors (MBR) combines activated sludge treatment with a membrane liquid-solid separation process. The membrane component utilises low pressure microfiltration or ultra filtration membranes and eliminates the need for clarifaction and tertiary filtration. The membranes are typically immersed in the aeration tank (however some applications utilise a separate membrane tank). One of the key benefits of a membrane bioreactor system is that it effectively overcomes the limitations associated with poor settling of sludge in conventional activated sludge (CAS) processes.

The technology permits bioreactor operation with considerably higher mixed liquor suspended solids (MLSS) concentration than CAS systems, which are limited by sludge settling. The process is typically operated at MLSS in the range of 8,000 -12,000 mg/L, while CAS are operated in the range of 2,000 - 3,000 mg/L. The elevated biomass concentration in the membrane bioreactor process allows for very effective removal of both soluble and particulate biodegradable materials at higher loading rates. Thus increased Sludge Retention Times

(SRTs)—usually exceeding 15 days—ensure complete nitrification even under extreme cold weather operating conditions.

The cost of building and operating a MBR is usually higher than conventional wastewater treatment, however, as the technology has become increasingly popular and has gained wider acceptance throughout the industry, the life-cycle costs have been steadily decreasing. As well, in developed urban areas where the footprint of the treatment plant is considered a limiting factor MBR Facilities can be considered a desirable option.

Secondary Sedimentation: The final step in the secondary treatment stage is to settle out the biological floc or filter material and produce sewage water containing very low levels of organic material and suspended matter.

Rotating Biological Contactors: Rotating biological contactors (RBCs) are mechanical secondary treatment systems, which are robust and capable of withstanding surges in organic load. RBC's were first installed in Germany in 1960 and have since been developed and refined into a reliable operating unit. The rotating disks support the growth of bacteria and microorganisms present in the sewage, which breakdown and stabilise organic pollutants. To be successful, microorganisms need both oxygen to live and food to grow. Oxygen is obtained from the atmosphere as the disks rotate. As the microorganisms grow, they build up on the media until they are sloughed off due to shear forces provided by the rotating discs in the sewage. Effluent from the RBC is then passed through final clarifiers where the microorganisms in suspension settle as a sludge. The sludge is withdrawn from the clarifier for further treatment.

Tertiary Treatment

Tertiary treatment provides a final stage to raise the effluent quality before it is discharged to the receiving environment (sea, river, lake, ground, etc.) More than one tertiary treatment process may be used at any treatment plant. If disinfection is practised, it is always the final process. It is also called "effluent polishing".

Filtration: Sand filtration removes much of the residual suspended matter. Filtration over activated carbon removes residual toxins.

Lagooning: Lagooning provides settlement and further biological improvement through storage in large man-made ponds or lagoons. These lagoons are highly aerobic and colonisation by native macrophytes, especially reeds, is often encouraged. Small filter feeding invertebrates such as Daphnia and species of Rotifera greatly assist in treatment by removing fine particulates.

Constructed Wetlands: Constructed wetlands include engineered reedbeds and a range of similar methodologies, all of which provide a high degree of aerobic biological improvement and can often be used instead of secondary treatment for small communities. One example is a small reedbed used to clean the drainage from the elephants' enclosure at Chester Zoo in England.

Waste Removal: Wastewater may contain high levels of the nutrients nitrogen and phosphorus. Excessive release to the environment can lead to a build up of nutrients, called eutrophication, which can in turn encourage the overgrowth of weeds, algae, and cyanobacteria (blue-green algae). This may cause an algal bloom, a rapid growth in the population of algae. The algae numbers are unsustainable and eventually most of them die. The decomposition of the algae by bacteria uses up so much of oxygen in the water that most or all of the animals die, which creates more organic matter for the bacteria to decompose. In addition to causing deoxygenation, some algal species produce toxins that contaminate drinking water supplies. Different treatment processes are required to remove nitrogen and phosphorus.

Nitrogen Removal: The removal of nitrogen is effected through the biological oxidation of nitrogen from ammonia (nitrification) to nitrate, followed by denitrification, the reduction of nitrate to nitrogen gas. Nitrogen gas is released to the atmosphere and thus removed from the water.

Nitrification itself is a two-step aerobic process, each step facilitated by a different type of bacteria. The oxidation of ammonia (NH_3) to nitrite (NO_2^-) is most often facillitated by *Nitrosomonas* spp. (nitroso=ammonium). Nitrite oxidation to nitrate (NO_3^-), though traditionally believed to be facilitated by *Nitrobacter* spp. (nitro=nitrite), is now known to be facilitated in the environment almost exclusively by *Nitrospira* spp.

Denitrification requires anoxic conditions to encourage the appropriate biological communities to form. It is facilitated by a wide diversity of bacteria. Sand filters, lagooning and reed beds can all be used to reduce nitrogen, but the activated sludge process (if designed well) can do the job the most easily. Since denitrification is the reduction of nitrate to dinitrogen gas, an electron donor is needed. This can be, depending on the wastewater, organic matter (from faeces), sulphide, or an added donor like methanol.

Sometimes the conversion of toxic ammonia to nitrate alone is referred to as tertiary treatment.

Phosphorus Removal: Phosphorus can be removed biologically in a process called enhanced biological phosphorus removal. In this process, specific bacteria, called polyphosphate accumulating organisms, are selectively enriched and accumulate large quantities of phosphorus within their cells (up to 20 per cent of their mass). When the biomass enriched in these bacteria is separated from the treated water, these biosolids have a high fertilizer value.

Phosphorus removal can also be achieved by chemical precipitation, usually with salts of iron (e.g. ferric chloride) or aluminium (e.g. alum). The resulting chemical sludge is difficult to handle and the added chemicals can be expensive. Despite this, chemical phosphorus removal requires significantly smaller equipment footprint than biological removal, is easier to operate and can be more reliable in areas that have wastewater compositions that make biological phosphorus removal difficult.

Disinfection

The purpose of disinfection in the treatment of wastewater is to substantially reduce the number of microorganisms in the water to be discharged back into the environment. The effectiveness of disinfection depends on the quality of the water being treated (e.g., cloudiness, pH, etc.), the type of disinfection being used, the disinfectant dosage (concentration and time), and other environmental variables. Cloudy water will be treated less successfully since solid matter can shield organisms, especially from ultraviolet light or if contact times are low. Generally, short contact times, low doses and high flows all militate against effective disinfection. Common methods of disinfection include ozone, chlorine, or ultraviolet light. Chloramine, which is used for drinking water, is not used in wastewater treatment because of its persistence.

Chlorination remains the most common form of wastewater disinfection in North America due to its low cost and long-term history of effectiveness. One disadvantage is that chlorination of residual organic material can generate chlorinated-organic compounds that may be carcinogenic or harmful to the environment. Residual chlorine or chloramines may also be capable of chlorinating organic material in the natural aquatic environment. Further, because residual chlorine is toxic to aquatic species, the treated effluent must also be chemically dechlorinated, adding to the complexity and cost of treatment.

Ultraviolet (UV) light can be used instead of chlorine, iodine, or other chemicals. Because no chemicals are used, the treated water's taste is more natural and pure as compared to other methods. UV radiation causes damage to the genetic structure of bacteria, viruses, and other pathogens, making them incapable of reproduction. The key disadvantages of UV disinfection are the need for frequent lamp maintenance and replacement and the need for a highly treated effluent to ensure that the target microorganisms are not shielded from the UV radiation (i.e., any solids present in the treated effluent may protect microorganisms from the UV light). In the United Kingdom, light is becoming the most common means of disinfection because of the concerns about the impacts of chlorine in chlorinating residual organics in the wastewater and in chlorinating organics in the receiving water. Edmonton, Alberta, Canada also uses UV light for its water treatment.

Ozone O_3 is generated by passing oxygen O_2 through a high voltage potential resulting in a third oxygen atom becoming attached and forming O_3. Ozone is very unstable and reactive and oxidises most organic material it comes in contact with, thereby destroying many pathogenic microorganisms. Ozone is considered to be safer than chlorine because, unlike chlorine which has to be stored on site (highly poisonous in the event of an accidental release), ozone is generated onsite as needed. Ozonation also produces fewer disinfection by-products than chlorination. A disadvantage of ozone disinfection is the high cost of the ozone generation equipment and the requirements for highly skilled operators.

Package Plants and Batch Reactors

In order to use less space, treat difficult waste, deal with intermittent flow or achieve higher environmental standards, a number of designs of hybrid treatment plants have been produced. Such plants often combine all or at least two stages of the three main treatment stages into one combined stage. In the UK, where a large number of sewage treatment plants serve small populations, package plants are a viable alternative to building discrete structures for each process stage.

The most advanced packaged treatment plant according to a study done by the University of California at Davis for the California State Water Resources Control Board for treating waste and nutrients (phosphorus and nitrogen) in one step economically is the USBF (Upflow Sludge Blanket Filter). The USBF process is a modification of the conventional activated sludge process that incorporates an anoxic selector zone and an upflow sludge blanket filtration clarifier all in one integrated bioreactor vessel. The treatment includes efficient reduction of BOD5 and TSS but also biological nutrient removal (BNR) by the processes of denitrification and "biological luxury uptake". The ensuing compact, modular system takes up less space and contains very few moving parts. The result is an efficient, highly affordable wastewater treatment plant with low maintenance and operating costs.

USBF technology has no inherent capacity limits and is used in a wide range of applications from subdivisions (Bay Point, FL) resorts (Sun Peaks, BC, Kicking Horse, BC) and municipalities (Pahrump, NV), to agricultural and industrial sites. Plants can be retrofitted and expanded from existing sites reducing capital costs. Since there are no mechanical parts and no chemicals needed, operations cost are much less than sequencing batch reactor and membrane bioreactor systems.

Another type of process which combines secondary treatment and settlement is the sequencing batch reactor (SBR). Typically, activated sludge is mixed with raw incoming sewage and mixed and aerated. The resultant mixture is then allowed to settle producing a high quality effluent. The settled sludge is run off and re-aerated before a proportion is returned to the head of the works. SBR plants are now being deployed in many parts of the world including North Liberty, Iowa, and Llanasa, North Wales.

The disadvantage of such processes is that precise control of timing, mixing and aeration is required. This precision is usually achieved by computer controls linked to many sensors in the plant. Such a complex, fragile system is unsuited to places where such controls may be unreliable, or poorly maintained, or where the power supply may be intermittent.

Package plants may be referred to as high charged or low charged. This refers to the way the biological load is processed.

In high charged systems, the biological stage is presented with a high organic load and the combined floc and organic material is then oxygenated for a few hours before being charged again with a new load. In the low charged system the biological stage contains a low organic load and is combined with floculate for a relatively long time.

Sludge Treatment and Disposal

The sludges accumulated in a wastewater treatment process must be treated and disposed of in a safe and effective manner. The purpose of digestion is to reduce the amount of organic matter and the number of disease-causing microorganisms present in the solids. The most common treatment options include anaerobic digestion, aerobic digestion, and composting.

The choice of a wastewater solid treatment method depends on the amount of solids generated and other site-specific conditions. However, in general, composting is most often applied to smaller-scale applications followed by aerobic digestion and then lastly anaerobic digestion for the larger-scale municipal applications.

Anaerobic Digestion

Anaerobic digestion is a bacterial process that is carried out in the absence of oxygen. The process can either be *thermophilic* digestion, in which sludge is fermented in tanks at a temperature of 55°C, or *mesophilic,* at a temperature of around 36°C. Though allowing shorter retention time (and thus smaller tanks), thermophilic digestion is more expensive in terms of energy consumption for heating the sludge.

One major feature of anaerobic digestion is the production of biogas, which can be used in generators for electricity production and/or in boilers for heating purposes.

Aerobic Digestion

Aerobic digestion is a bacterial process occurring in the presence of oxygen. Under aerobic conditions, bacteria rapidly consume organic matter and convert it into carbon dioxide. The operating costs are characteristically much greater than for anaerobic digestion because of the energy costs needed to add oxygen to the process.

Composting

Composting is also an aerobic process that involves mixing the wastewater solids with sources of carbon such as sawdust, straw or wood chips. In the presence of oxygen, bacteria digest both the wastewater solids and the added carbon source and, in doing so, produce a large amount of heat.

Thermal Depolymerisation

Thermal depolymerisation uses hydrous pyrolysis to convert reduced complex organics to oil.

Sludge Disposal

When a liquid sludge is produced, further treatment may be required to make it suitable for final disposal. Typically, sludges are thickened (dewatered) to reduce the volumes transported off-site for disposal. There is no process which completely eliminates the need to dispose of biosolids. There is, however, an additional step some cities are taking to superheat the wastewater sludge and convert it into small pelletised granules that are high in nitrogen and other organic materials. This product is then sold to local farmers and turf farms as a soil amendment or fertilizer, reducing the amount of space required to dispose of sludge in landfills.

Treatment in the Receiving Environment

Many processes in a wastewater treatment plant are designed to mimic the natural treatment processes that occur in the environment, whether that environment is a natural water body or the ground. If not overloaded, bacteria in the environment will consume organic contaminants, although this will reduce the levels of oxygen in the water and may significantly change the overall ecology of the receiving water.

Native bacterial populations feed on the organic contaminants, and the numbers of disease-causing microorganisms are reduced by natural environmental conditions such as predation exposure to ultraviolet radiation, for example. Consequently, in cases where the receiving environment provides a high level of dilution, a high degree of wastewater treatment may not be required. However, recent evidence has demonstrated that very low levels of certain contaminants in wastewater, including hormones (from animal husbandry and residue from human birth control pills) and synthetic materials such as phthalates that mimic hormones in their action, can have an unpredictable adverse impact on the natural biota and potentially on humans if the water is reused for drinking water. In the US and EU, uncontrolled discharges of wastewater to the environment are not permitted under law, and strict water quality requirements are to be met. A significant threat in the coming decades will be the increasing uncontrolled discharges of wastewater within rapidly developing countries.

Sewage Treatment in Developing Countries

There are few reliable figures on the share of the wastewater collected in sewers that is being treated in the world. In many developing countries the bulk of domestic and industrial wastewater is discharged without any treatment or after primary treatment only. In Latin America about 15 per cent of collected wastewater passes through treatment plants (with varying levels of actual treatment). In Venezuela, a below average country in South America with respect to wastewater treatment, 97 per cent of the country's sewage is discharged raw into the environment. Even a highly industrialised country such as the People's Republic of China discharges about 55 per cent of all sewage without treatment of any type. In a relatively developed Middle Eastern country such as Iran, Tehran's majority of population has totally untreated sewage injected to the city's groundwater. Most of sub-Saharan Africa is without wastewater treatment.

Water utilities in developing countries are chronically

underfunded because of low water tariffs, the inexistence of sanitation tariffs in many cases, low billing efficiency (i.e. many users that are billed do not pay) and poor operational efficiency (i.e. there are overly high levels of staff, there are high physical losses, and many users have illegal connections and are thus not being billed). In addition, wastewater treatment typically is the process within the utility that receives the least attention, partly because enforcement of environmental standards is poor. As a result of all these factors, operation and maintenance of many wastewater treatment plants is poor. This is evidenced by the frequent breakdown of equipment, shutdown of electrically operated equipment due to power outages or to reduce costs, and sedimentation due to lack of sludge removal. Developing countries as diverse as Egypt, Algeria, China or Colombia have invested substantial sums in wastewater treatment without achieving a significant impact in terms of environmental improvement. Even if wastewater treatment plants are properly operating, it can be argued that the environmental impact is limited in cases where the assimilative capacity of the receiving waters (ocean with strong currents or large rivers) is high, as it is often the case.

Waterborne diseases that are prevalent in developing countries, such as diarrhea, typhus and cholera, are caused primarily by poor hygiene practices and the absence of improved household sanitation facilities. The public health impact of the discharge of untreated wastewater is comparatively much lower. Hygiene promotion, on-site sanitation and low-cost sanitation thus are likely to have a much greater impact on public health than wastewater treatment. Given the scarcity of financial resources in developing countries and the poor track record of wastewater treatment plants, it could thus be argued that investments should first be undertaken to evacuate wastewater from human settlements and to promote good hygiene practices. Only once this has been achieved, substantial funds should be invested in wastewater treatment plants.

10

Solid Organic Waste Processing

Rapid proliferation of industries, increased human population, urbanisation and misuse and abuse of the fragile environment have led to increased accumulation of wastes. All these processes have significantly reduced the availability of fertile agricultural lands. Large-scale environmental degradation due to air, water and soil pollution have also been observed in last few decades due to these anthropogenic activities. Further, to enhance food production, for sustaining large human population, protect the high yielding crops from pest and pathogen attacks, large-scale use of synthetic chemical fertilizer, pesticides, herbicides, nematicides and fungicides have been used allover the world. Despite enhanced crop yield and food security all these practices have caused irreparable ecological damages. Ecologists all around have strongly advocated in favour of organic farming. Huge quantity of urban and rural organic wastes generated are now being processed and utilised to boost agricultural production, for fuel generation and energy recovery.

Generation of Organic Waste

Due to large population and high rate of consumption, considerable quantity of organic waste are generated in the cities. Accumulation of these wastes often creates environmental problems, besides damaging the aesthetics of the urban environment. Hence for maintenance of a healthy and sustainable environment, large-scale utilisation of these wastes is a prerequisite.

The major sources of solid organic wastes in cities are household waste, institutional refuse, street sweeping waste, ashes and residues, animal wastes, dead animals, etc. Solid waste management has been practised globally and developed nations such as USA, England, Canada, Germany, Ireland, Portugal and many others, have enacted laws for proper disposal and recycling of municipal solid waste.

In, India, average per capita generation of solid waste is about 300 to 600 gms\per day. The total municipal solid waste generated by 23 metros in India, has been estimated to be 30,058 tones per day. Besides the major cities, fast growing cities such as Bhubaneswar with a present population of over 7 lakhs, generate more than 600 tones of municipal solid waste everyday (Panda et.al.2003).

Besides the urban municipality wastes, many large and medium industries such as pulp and papermills, cotton mills, food processing industries, tanneries etc produce various kinds of organic wastes in considerable quantity.

It has long been realised that technologies should be available for large-scale processing of the waste generated in urban areas and the industries so as to convert them into useful products.

Earthworm Biodiversity

Earthworms are the dominant macrofauna in the soil ecosystem. These invertebrates along with the microbial communities carryout the decomposition of organic waste materials, thus maintaining the nutrient flux in the soil. Earthworms play tremendous role in the development and conservation of soil. Although the usefulness of the worms were realised since late nineteenth century, systematic and scientific methods for their utilisation for environmental amelioration was only realised in the mid twentieth century. At present vermitechnology has become an established, eco friendly organic waste processing method all over the globe.

Many workers have studied mechanisms of processing of organic waste by using various species of earthworms. Reports are available on earthworm decomposition of organic waste such as pig and cattle solids, chicken wastes, horse manure, rabbit dropping, crop residues and city garbage.

Stephenson (1923) was probably the first to report earthworm fauna in the Indian subcontinent. Gates (1972) has reported earthworm species in Indo-Gangetic plains. Julka and Senapati (1987) reported about 30 different species of earthworms from the state of Orissa. Bhadauria and Ramakrishnan (1989) studied the population dynamics of earthworm in agrcecosystems. Ismail *et al.* (1990), Bano and Kale (1991) have reported earthworm fauna from southern India.

Lampito mauritii is perhaps the most widely distributed earthworms in India. Besides, this species *Polypheretima elongata, Perionyx excavatus, Dichogaster bolaui,* show wide distribution in India. Further the species which are abundant regionally are *Perionyx milardii* and *Octochaetona surensis.* The factors which largely influence the distribution and abundance of earthworms are the soil pH, organic matter, moisture percentage, temperature range, agricultural practices, cropping patterns, etc. (Kale and Krishnamoorthy, 1978; Julka and Senapati, 1987; Choudhuri and Bhattacharjee, 1999).

Earthworms are generally abundant in traditional organic agroecosystems, where organic manure input is adequate (Kale, 1997).

Earthworms Agents of Biodegradation

Earthworms selectively feed on decomposing organic matter and defecate partially digested material on soil surface, as mucous coated castings. Release of CO_2 through respiration, production of mucous and nitrogen rich excrement's enhance the level of nitrogen in soil and lowers the C/N ratio. Worm castings show significantly higher concentrations of total nitrogen, organic matter, nitrate, nitrogen, phosphrous, potassium, sodium and magnesium. (Dash and Patra, 1979; Reddy, 1983; Bhadauria and Ramakrishnan, 1989, 1996). Certain earthworm casts are rich in soluble calcium and carbonates. Earthworms are called secondary decomposers since they catalyse the process of organic residue fragmentation by increasing the surface area exposed to microflora. The earthworm soil microbe associations and interactions varies with soil types and management conditions (Edwards and Bohlen, 1996).

Culturing the Worms

Earthworm culture is generally done indoor in humid places and in shady areas to avoid direct sunlight and heavy down pour. For culture cement tanks, wooden boxes or plastic trays of the size 1m x 1m x 0.5 m could be used which can accommodate 1000 worms. The culture pits should be covered suitably to protect the worms from predator and prevent them from escaping during night time. Saw dust or husk, fine sand or garden soil may be used as bedding materials. 3 cms of each of the material is to be spread in the culture pits. Water is to be sprinkled on the bed to get a moisture level of nearly 50 per cent. After the heed preparation, the worms can be induced into the pits.

Cattle, sheep, horse, pigs dung and droppings of poultry birds are ideal feed for the worms. Further wheat bran and vegetable waste when added to the dung in 10:1 ratio would enhance biomass production of the worm considerably (Bano and Kale, 1984). The feed is to be provided in small quantity at a time and be compensated as and when required in the culture pits.

Though the methods of culturing worms may vary from place to place there are certain prerequisites for feasible methods of vermiculture. A cultured worm must be able to adapt to various substrates, must have high multiplication rate and grow faster under controlled condition. Some of the ideal species for vermiculture such as—*Lumbricus rubellus, Perionyx excavatus, Eisenia foetida, Dichogaster bolaui, Eudrilus eugeniae* have been found suitable for large-scale vermicomposting. *Lumbricus rubelus* under favourable moisture and temperature conditions lays an egg capsule (cocoon) every 7 to 10 days. Each capsule containining 7 embryos. *Perimyx excavatus* grows very fast and produces 7 to 31 young worms per month. About 2000 mature worms if scientifically cultured could produce more than one million worms in a year.

Resource Recovery from Organic Waste through Vermitechnology

Vermiculture and vermicomposting has of late emerged as an ideal and eco-friendly method for organic waste management. Vermiculture helps in improving the physical qualities of the waste material and reducing the smell and putrescent potential of the decomposing waste. Appelhof (1974, 1980), Gaddii (1976), Dash (1978), Hartenstein *et al.* (1979), Edwards and Bohlen (1995) have indicated that earthworms can successfully process household garbage, municipality waste and organic industrial waste which could be transformed into valuable manure. It has also been reported that the vermicompost produced from organic wastes are rich in nutrient which become readily available to plants and hence could be utilised in agriculture to improve soil structure and fertility.

Industrial Waste Processing

Earthworms could be successfully utilised for processing organic solid waste from various industries such as paper mills, oil extraction plants, organochemical industries, diary industries, tanneries, treatment of hospital wastes and waste from enzyme production units. Useful biofertilizer and protein biomass can be produced by application of vermitechnology in these industries. Earthworms in general can suitably process industrial waste, which has a pH range between 5.5 to 8.5 and solid content of 40 per cent to 50 per cent. Where the percentage of solid content is less it could be compensated by fly ash, lime sludge, paper clipping, canteen and kitchen waste or even toilet wastes. Other components which could increase solid content of the waste are cellulose wastes such as discarded newspaper, saw dust, fresh on dry cattle dung or even green foliage.

Aclimatisation of the suitable species for vermitechnology to the waste is essential at the initial stages of vermicomposting. This may take about 4 to 8 months later in the retention time may drop even to 40 to 45 days. Sometimes inoculation of decomposing micro organisms helps faster degradation of the organic waste and reduces the time of vermicomposting. Especially certain cellulose degrading fungi, *Actinomycetes* and bacterial strains have proved to enhance the process of decomposition. It has been reported earlier that *Aspergillus* and *Cellulomonas* with earthworms degrade paper mill effluents, quite efficiently.

Conclusion

Considering the fast deteriorating urban environment, sustainable and eco-friendly waste processing technologies such as vermitechnology seem to have tremendous potential. Besides, utilising various exotic varieties of earthworm available for vermicomposting, research and development in the suitability of native species of worms in waste management is the need of the hour.

Biodetoxification of Leather Industry

The major problem regarding the industrial effluent is their toxic effect towards the environment when buried into soil or diffused in to air or dumped into water. The toxic chemicals of many types enter water resources are mostly from industries. These chemicals include mineral acids, inorganic acids, metals and metallic compounds, trace elements, cyanides, sulphates, nitrites, organometallic compounds etc (Abha srivastava and Pathan, 1998).

Regarding leather industry, about 130 different chemicals are used in the manufacturing process, which ranges from common salt to very expensive chromium sulphate. The leather industry waste which are oxygen demanding, leads to depletion of dissolved oxygen in the aquatic environment. The dissolved oxygen level should be normally at least 5mg/l for the survival of aquatic life (Rollinsom.*et al.*, 1973). Microorganisms are very important in geochemical cycling of organic and inorganic nutrients and biodegradation and biotransformation of pollutant in the environment (Venitt *et al.*, 1974). By using bacterial strain isolated from activated sludge of distillery waste water treatment plant, COD removal and colour removal is done successfully (Nanjunda swamy *et al.*, 1998). The most organic compounds are vulnerable to microbial attack (Ajaybabu *et al.*, 1999).

Many bacterial strains *Desulfovibrio, Enterobacter, Alcaligens, Dinococcus* has been described for their ability to reduce toxicity both aerobically and anaerobically. Among these *Pseudomonas sp.* can detoxify the leather industry effluent containing predominant organic and inorganic compounds effectively. The presence study focuses the Biodetoxification of leather industry effluent using free and immobilised biocatalyst (*Pseudomonas* cells). The parameters analysed are physicochemical analysis (pH, TDS, TSS), inorganic analysis (Sulphate and chloride) and organic analysis (DO, BOD, COD, oil and grease) and the survival efficiency of fishes (*Lepidocephalicthys thermalis*).

Materials and Methods

Collection of Samples: The sample was collected from the leather industry. The sample bottle were autoclaved at 15 lbs,

121°C for 15 min. The lid of the bottle was removed at the collection site by taking the bottle into the water in inverted position up to the depth of 1-2 feet below the water surface. It was immediately closed and taken out.

Isolation of Organism

Pseudomonas sp. was isolated by serially diluting the leather industry effluent sample up to 10^{-6} dilution and plated on the selective medium (cetrimide agar).

Immobilizaton of Culture

The isolated culture of *Pseudomonas sp.* was grown in king's broth at 37° C for 24 hrs. Pellet out the cells and mixed with sodium alginate solution and extruded drop wise to 4 per cent filtered calcium chloride solution. They were stabilised and used for inoculation.

Inoculation of Sample

The inoculation was done in two modes. In the first sample was inoculated with 10 per cent of free cell culture of *Pseudomonas sp.*

In the second type the sample was inoculated with 10 per cent of immobilised cells of *Pseudomonas sp.* Both samples are incubated at two conditions, one set with aeration and other set without aeration for 5 days at 20°C.

Parameters Analysed

Physicochemical analysis (colour, odour, pH,TDS,TSS,TS) Inorganic analysis (sulphate, chloride) & Organic analysis(DO,BOD,COD, oil & grease) were analysed (APHA,AWWA and WPCF,1975).

Result and Discussion

The physicochemical, inorganic and organic parameters were experimented for the leather industry effluent under unaerated and aerated condition.

Table: Parameters analysed under unaerated condition

S. no	*Parameters*	*Raw effluent (control)*	*Effluent + Free cells*	*Effluent + Immobilised cells*
1.	Colour	Dark Brown	Pale yellow	*Colourless*
2.	Odour	Undesirable	Culture smell	*No undesirable smell*
3.	PH	8.1	8.0	*7.8*
4.	TDS(mg/l)	4421	4290	*4170*
5.	TSS(mg/l)	170	150	*100*
6.	TS(mg/l)	210	180	*120*
7.	Sulphate	810	760	*680*
8.	Chloride	2411	2370	*2200*
9.	DO	-	-	*1.60*
10.	BOD	72	55	*35*
11.	COD	370	320	*240*
12.	Oil and grease	20	11	10

Table: Parameters analysed under aerated condition

S. no	*Parameters*	*Raw effluent (control)*	*Effluent + Free cells*	*Effluent + Immobilised cells*
1.	Colour	Light Brown	Pale yellow	*Pale brown*
2.	Odour	Undesirable	Culture smell	*No undesirable smell*
3.	PH	8.0	7.9	7.7
4.	TDS(mg/l)	4400	4280	4110
5.	TSS(mg/l)	164	130	80
6.	TS(mg/l)	192	150	100
7.	Sulphate	795	742	667
8.	Chloride	2390	2310	2150
9.	DO	–	1.40	3.10
10.	BOD	61	46	24
11.	COD	350	300	210
12.	Oil and grease	18	17	9

The BOD reveals the amount of organic and inorganic pollutant load in the effluent. BOD is the measure of amount of oxygen used in the respiratory process of the organism in oxidising organic content in the effluent. The characteristics of untreated waste from industries are very strong and they have high BOD and suspended solids. The treatment would be better, if microbiological treatment system is used (Subbaiah.1981). Biological treatment is an alternative method to the physicochemical method and it is more advantageous due to its eco-friendly and cost-effective nature (Balasubramaniyan. 2000).

Immobilised cells provide multi-enzyme system of degradation. The important feature in cell immobilisation technology is that, the enzymes are active and stable for long period of time. It keeps with in the cellular domain together with all cell constituents, whether the cells are dead or viable but in resting stage. Degradation of phenol can be successfully done by using immobilised cells of Pseudomonas (Keshar *et al.*, 1991).

The recycling of immobilised beads can be done up to six cycles with the degradation efficiency of 73 per cent (Ramanathan *et al.*, 1996). The efficiency of Biodetoxification was determined by the survival efficiency of fishes *(Lepidocephalicthys thermalis)* for 48 hours in the raw and treated sample. The assurance of decrease in toxicity of the effluent can be determined only by the comparative survival ability of fishes on the untreated and treated samples. (Anandapandian *et al.*, 2002)

Table: Survival efficiency of fishes under unaerated condition

S.No. Condition exposed	*100%*	*75%*	*50%*	*25%*
1. Raw effluent	-	-	-	-
2. Effluent + Free cell	-	-	2	3
3. *Effluent + Immobilised cells*	-	1	2	3

Table: Survival efficiency of fishes under; condition aerated

S.No. Condition exposed	*100%*	*75%*	*50%*	*25%*
1. Raw effluent	-	-	-	-
2. Effluent + Free cell	-	-	3	4
3. *Effluent + Immobilised cells*	-	2	4	5

The sample treated with immobilised cells under aerated condition shows low mortality rate of fishes due to the fixing of oxygen and effective degradation of organic and inorganic compounds in the effluent. So this biological treatment can be used for the development of effluent treatment plant.

Phytoremediation of Mercury Contaminated Waste from a Chlor-Alkali Factory

A delicate balance exists between man and his environment which has been seriously disturbed by human activities. Pollution, now a days has become a watch word and anything which tampers with the ecosystem, is a matter of concern. Thousand of toxic contaminants have found their way into environment due to human interference with nature. Though there is certain amount of inherent antagonism between development and environment still then we cannot forego development. The wastes released from various industries has received much attention due to its toxicity and infectious nature.

The conventional processes used for effluent treatment are precipitation, hydroxides / sulphides, oxidation / reduction, and ion exchange. These processes are expensive and not eco-friendly (Gadd & Griffiths, 1978; Volesky, 1990). That is why a biological means of pollution control is most favoured today and phytoremediation can be suggested as an effective method.

The heavy metal, along with other pollutants can be accumulated biological (Mishra, 1987; Srivastava & Aditya; Dash 2002) and subsequently transferred with biomagnification in different trophic levels in a food chain (Boudou *et al* 1979; Kirobal *et al.*, 1983) Siegal *et al.*, 1977, Temmerman *et al.*, 1986, Dash 2002, reported selectivity in the mercury accumulation by plants.

Case Study with Jayashree Chemicals a Chlor-alkali Factory

On the bank of the river Rushikulya is situated Ganjam town, about 4km from the shore of the Bay of Bengal where the river meets the sea. Ganjam town perceived the threat of mercury pollution with the establishment of the chlor-alkali factory—M/s. Jayashree Chemicals, Ganjam in 1962 with production of caustic soda and chlorine gas. The factory started its production in 1967 with an installed capacity 16.5 thousand tones per year and releases 2.7 tonnes of mercury to the nearby locality of the Industrial unit. The initial charge of the mercury cathode was 54 tonnes of mercury every year there after the capacity of the factory was increased and it started indenting mercury 3 tonnes more per year. The factory releases 50,000 gallons of mercury contaminated effluent to the Rushikulya river every year. Exhaust air releases mercury directly from the cell room through ventilators. Waste soil after electrolysis of the brine water, which is dumped by the side of the fluent channel is heavily contaminated with mercury, sodium and chloride. An analysis of the wastes and sample around the factory indicate high contamination of the area with the heavy metal mercury.

Along with contamination the vegetation, the cultivated crops and fishes around the chlor-alkali factory, are also exposed to the threat of mercury pollution from the industry. It has been reported that the fishes in the river Rushikulya and Bay of Bengal are contaminated with an approximately amount of mercury (Shaw *et al.*, 1985). The local crop field are also effected with the pollutants released by the industry. An analysis of the water from the local crop field (Nanda *et al.*, 1986b) indicates contamination of the crop fields effecting adversely to its production in near future.

Accumulation of Mercury by Sesbania Aculeata and Oryza sativa

Sesbania aculeata and is an erect, semi- woody plant having 4-5ft height. The plant is having tap root systems with root nodules for fixation of nitrogen. The plant has weak prickles over the stem and leaf rachis. The leaves are unipinnately compound which are 6 to 12 inches long with 20 to 10 pairs of linear leaflets. The leaflets are obtuse, apiculate and glabrous in nature. The stipules are minute and caducous. The flowers are yellowish-white in colour which are .3 to .5 inches long. The pods are slender in nature and are 6-9 inches long The fruits are beaked. The plant is long known as a biefertilizer for rice cultivation in the state of Orissa. It is cultivated as a crop rotation plant. After the Khariff harvest in non- irrigated areas there is a usual practice of cultivating *Sesbania* sp. during *Ravi* season. *Sesbania* is a green manuring agent, which fixes atmospheric nitrogen by the symbiotic *Rhizobium* present in the root forming root nodules. The nodules makes the soil rich with nitrogen and again at the advent of rainy season the plants are cut and allowed to decompose thereby releasing organic carbon and additional nitrogen to the soil. Available information's indicate *Sesbania* release 30 to 50kg of nitrogen per hectare.

Oryza sativa is an erect, herb having 5ft height. The plant has fibrous root, linear leaf with sheathing leaf base, parallel venation, spikelet inflorescence. The perianth is represented by lodicules. Ovary is monocarpellary, Superior, Unilocular, Single Ovule, basal placentation. Fruit is caryopsis. The rice is the main staple food of India.

Reports available are very few pertaining to the accumulation and effects of mercury of higher plants. Therefore an attempt was made in this investigation to determine accumulation of mercury by Sesbania and Rice.

Materials and Method

Analysis of the Waste Soil: The waste soil of the chlor-alkali factory is periodically removed and dumped as heaps without any chemical processing for removal of its pollutants before they are released. The waste soil were collected in gunny bags and kept under air in field condition for 48 hours to decrease its moisture level. It was then powered, sieved and subjected to analysis for various important constituents. pH of the gray coloured soil measured to be 9.20 ± 0.03. It contained 0.750mg of mercury per kg of the waste soil as analysed by cold vapour atomic absorption technique. Sodium content of the waste analysed by flame photometer was 5.9kg[1] and chloride content was 13g kg^{-1} of the waste. Potassium and phosphate contents of the waste soil were 62 and 72 mg per kg respectively.

Preparation of Culture Pots (Sesbania Aculeata)

The dried and powdered waste soil was used for preparation of culture pots with garden soil in order to study the effect of the waste soil combination on *Sesbania* and effect of *Sesbania* on waste soil combinations. At an interval of 10 per cent, varying concentrations (10% to 60%) of waste soil was prepared with normal garden soil (pH 6.75 ± 0.05) with final weight 4Kg. They were kept in earthen pots of equal size to use as culture medium for *Sesbania*. Pots were kept in field conditions. They were watered equally. The soil water mixture was allowed to settle for two days. Seeds of *Sesbania* were sown in the pots containing varying contaminated waste soil combinations. The pots were arranged in triplicate and various parameters were observed from the 7th day of after germination upto 35 days at an interval of 7 days.

Preparation of Culture Pots (Oryza sativa)

The dried and powdered waste soil was passed through a fine mesh and was used for the preparation of culture pots. Varying concentrations of the waste soil was prepared with normal garden soil (pH 6.44 ± 0.02) ranging from 2.5 per cent to 17.5 per cent at an interval of 2.5 per cent. Final weight of the soil combinations prepared was 4kg. They were kept in earthen pots of equal size in triplicate to use as the medium to culture rice seedlings. The pots were watered equally from tap water at the rate of soil: water = 1: The waste soil combinations and water were mixed thoroughly and allowed to settle. The level of water in all pots was marked with permanent paints and water level in the waste soil combinations was maintained daily. The experiment was carried out in field condition during normal Kharif season of rice transplantation. It is the period of traditional Kharif season for rice cultivation, usually availed by local farmers. After the soil suspensions settled, the supernatant of water of the solution was taken for determination of soil pH. A control set with only garden soil was maintained for comparison.

Methods of Analysis

After 7,14,21,28,35 days of transplantation the *Sesbania* plants were removed washed thoroughly, blotted and analysed for mercury accumulation. For Rice after 15 days and 30 days of transplantation on plants were removed and analaysed in the same way. The values obtained were subjected to statistical analysis (Mishra & Mishra, 1993) and the levels of significance were determined.

Result

Mercury Accumulation by Sesbania: Mercury accumulation by *Sesbania* from the culture medium was found to be both concentration and time dependent. It increased with increase in concentration of waste soil from 10 per cent to 60 per cent and with increased in time period from 7 days to 35 days.

Mercury Accumulation by Root

Table given below depicts the mercury accumulation by *Sesbania* root 7 days and 14days after germination period mercury accumulation by the root increased from 4.86 ± 0.06 μg g^{-1} fresh weight in 60 per cent and from 7.11 ± 0.11 μg g^{-1} fresh weight in 10 per cent to 18.27 ± 0.07 ug g^{-1} fresh weight in 60 per cent waste soil from the culture pots. Highly significant ($p < 0.01$) positive correlation ($r = 0.994$ & 0.991) were obtained by mercury accumulation and waste soil concentrations.

With increase in germination period to 21 days and 28 days mercury accumulation increased from 9.16 ± 0.06 μg g^{-1} fresh weight in 10 per cent to 21.34 ± 0.04 μg g^{-1} fresh weight in 60 per cent and from 11.00 ± 0.05 μg g^{-1} weight in 10 per cent to 24.00 ± 0.05 μg g^{-1} fresh weight in 60 per cent soil respectively.

Table: Mercury Accumulation by Root Fw) of Sesbania Percentage of solid waste

Days	10	20	30	40	50	60
7	4.86 ±0.06	6.22 ±0.02	9.10 ±0.10	10.68 ±0.08	12.43 + 0.03	13.85 ±0.05
14	7.11 ±0.11	10.65 ±0.05	12.42 ±0.10	14.18 ±0.06	17.00 ±0.05	18.27 ±0.07
21	9.16 ±0.06	13.00 ±0.50	14.45 ±0.95	16.25 ±0.05	19.27 ±0.07	21.34 ±0.04
28	11.00 ±0.05	15.35 ±0.05	18.10 0.10	19.65 ±0.05	22.22 ±0.02	24.00 ±0.05
35	12.85 ±0.05	17.05 ±0.05	21.00 ±0.85	22.44 ±0.04	24.16 ±0.06	25.61 ±0.06

Correlation between mercury accumulated by *Sesbania* from the waste soil combinations after 21 and 28 days of culture with waste soil concentrations were highly significant ($r = 0.990$ & 0.982, $p < 0.01$ respectively). In 35 days after germination period mercury uptake by Sesbania increased from 12.85 ± 0.05 μg g^{-1} fresh weight in 10 per cent to 25.61 ± 0.06 μg g^{-1} fresh weight in 60 per cent. Highly significant correlation was obtained between mercury accumulation by the root and per cent waste soil combinations ($p < 0.01$, $r = 0.971$).

Mercury Accumulation by Shoot

Table given below depicts the mercury accumulation by *Sesbania* Shoot. 7 days and 14 days after germination period mercury accumulation by the shoot increased from 2.46 ± 0.06 μg g^{-1} fresh weight in 10 per cent to 7.65 ± 0.05ug^{-1} fresh weight in 60 per cent and from 3.88 ± 0.07 μg g^{-1} fresh weight in 10 per cent 8.22 ± 0.20 μg g^{1} fresh weight in 60 per cent waste soil from the culture pots. Highly significant ($p < 0.01$) positive correlation ($r = 0.996$ & 0.986) were obtained by mercury accumulation and waste soil concentration. With increase in germination period to 21 days and 28 days mercury accumulation increased from 5.10 ± 0.05 fag g^{-1} fresh weight in 10 per cent to 9.90 ± 0.10 fag g^{-1} fresh weight in 60 per cent and from 6.20 ± 0.10 µg g^{-1} fresh weight in 10 per cent 10.45 ± 0.15 µg g^{-1} fresh weight 60 per cent soil respectively. Correlation between mercury accumulated by *Sesbania* from the waste soil combinations after 21 and 28 days of culture with waste soil concentrations were highly significant (r = 0.996 and 0.818, $p < 0.01$, respectively). In 35 days after germination period mercury uptake by *Sesbania* increased from 7.00 ± 0.09 µg g^{-1} fresh weight in 10 per cent to 11.15 ± 0.05 µg g^{-1} fresh weight in 60 per cent. Highly significant correlation was obtained between mercury accumulation by the shoot and per cent waste soil combinations ($p < 0.01$, r=0.984).

Mercury Accumulation by Rice

Table: Mercury Accumulation by Shoot (µg/gm/ Fw) of Sesbania Percentage of solid waste

Days	*10*	*20*	*30*	*40*	*50*	*60*
7	2.46 ±0.06	3.11 ±0.02	4.43 ±0.13	5.65 ±0.15	6.50 ±0.10	7.65 ±0.15
14	3.88 ±0.07	4.34 ±0.09	5.59 ±0.09	6.90 ±0.10	7.85 ±0.04	8.22 ±0.20
21	5.10 ±0.05	5.88 ±0.09	7,10 ±0.05	8.00 ±0.15	9.25 ±0.12	9.90 ±0.10
28	6.20 ±0.10	7.18 ±0.05	8.25 ±0.05	9.10 ±0.03	10.00 ±0.11	10.45 ±0.15
35	7.00 ±0.09	8.29 ±0.09	9.11 ±0.06	10.10 ±0.10	10.85 ±0.13	11.15 ±0.05

Mercury accumulation by *Oryza* from culture medium was found to be both time and concentration dependent.

Mercury Accumulation in Root

Mercury accumulation by root of rice seedlings was found to be dependant on both waste soil percentage and time. Mercury accumulation was found to increase with increase in waste soil from 2.5 per cent to 17.5 per cent and also time period after transplantation from 15 days to 30 days. In the control set with only garden soil mercury accumulation was not reported. After 15 days transplantation of the rice seedlings mercury uptake increased from 5.10 ± 0.38μg g^{-1} fresh weight 2.5 per cent to 14.66 ± 0.25 μg g^{-1} fresh weight in 17.5 per cent waste soil combination.

A highly significant positive co-relation ($P < 0.001$) was obtained between percentage of waste soil and mercury uptake by root (r=+0.992). 30 days transplantation of rice seedlings, mercury accumulation increased from 9.43 ± 0.38 μg g^{-1} in 2.5 per cent to 24.26 ± 0.26 μg g^{1} fresh weight in 17.5 per cent waste soil combination. The positive correlation (r = +0.998) between waste soil combinations and mercury uptake by root was highly significant ($P < 0.001$).

Mercury Accumulation by Shoot

Table given below depicts accumulation of mercury in shoot of rice seedlings. The accumulation of mercury in shoot was dependent on amount of waste soil and time period. It increases with increasing waste soil from 2.5 per cent to 17.5 per cent and from 15 days after transplantation to 30 days after transplantation. Highly significant correlations were observed between mercury uptake shoot and varying waste soil combinations.

After 15 days transplantation of the rice seedlings mercury accumulation in shoot increased from 2.45 ± 0.18 μg g^{-1} fresh weight in 2.5 per cent to 9.82 ± 0.75 μg g^{-1} fresh in 17.5 per cent waste sol combination.

The correlation between mercury uptake in shoot and per cent waste soil combinations were positive and highly significant (r=+0.991, $P < 0.001$). 30 days after transplantation, mercury accumulation increased from 3.70 ± 0.22 μg in 2.5 per cent to 11.11 ± 0.14 μg in 17.5 per cent waste soil combination per g fresh weight of shoot. A highly significant ($P < 0.001$) positive correlation (r= + 0.996) was obtained between mercury accumulation and waste soil combinations.

Table: Bivariate Correlation Coefficient Values for Mercury Uptake by root and shoot With Percentage of Waste soil of *Sesbania*

Days	*Variable*	*Correlation Values* ®	*Significance level (pl-)*	*Degrees of freedom*
7	Root	-0.994	0.01	5
	Shoot	0.996	0.01	5
14	Root	-0.991	0.01	5
	Shoot	-0,986	0.01	5
21	Root	-0.990	0.01	5
	Shoot	-0.996	0.01	5
28	Root	-0.982	0.01	5
	Shoot	-0.818	0.01	5
35	Root	-0.971	0.01	5
	Shoot	-0.984	0.01	5

Discussion

The accumulation and biological conversion (methylation) of mercury by producer organisms in aquatic and terrestrial ecosystems (Srebocan *et al.*,1981) is a matter of great concern. Roger (1976) reported methylation of divalent mercury in agricultural soil, degree of methylation directly proportional to the concentration of mercury in the soil to the exposure time. Eventually, Mercury is transferred to consumer organisms in the food chain.

The experiment discussed in this chapter shows that the accumulation of mercury by rice seedlings and *Sesbania* is dependent on both time and waste soil concentration. Herbage, vegetables, and crops accumulated mercury from mercurial herbicides (Maclean, 1974) and from industrial sources (Fleming and Parie, 1976); Mercury was accumulated from a chlor-alkali: factory by moss (Wallin, 1976). In these cases, accumulation was dependent on the availability of mercury to the organism (Shaw *et al.,* 1987). The mercury accumulated by rice seedlings and *Sesbania* in the present investigation is thus subject to distribution and transfer in the food chain.

The general use of plants to remediate environmental media *in situ*. It includes rhizofiltration phytoextraction, phytotransformation, phytostimulation and phyto stabilisation (Thakur, 2006)

In the last few years interest has been developed for phytoremediation of not only naturally land disturbed and but also of land contaminated or polluted with industrial discharges. However, serious attempts are required to make this system exploitable. Plants used in bioremediation can be genetically manipulated to enhance the efficiency and specificity of this function.

Table: Mercury Accumulation (µG G-¹ Fresh Weight) From Varying Waste Soil Combinations by Root and Shoot in Different Time Intervals

	Days	Percentage of Solid waste						
		2.50	5.00	7.50	10.00	12.50	15.00	17.50
Root	15	5.10	7.22	9.58	10.55	12.00	13.32	14.66
		±0.38	±0.62	±0.48	±0.39	±0.36	±0.19	±0.25
	30	9.43	11.66	14.75	16.21	19.05	22.00	24.26
		±0.38	±0.52	±0.43	±0.23	±0.27	±0.39	±0.26
Shoot	15	2.45	3.18	4.51	5.66	6.55	7.69	9.82
		±0.18	±0.26	±0.41	±0.50	±0.55	±0.71	±0.75
	30	3.70	4.45	5.90	6.95	8.05	9.60	11.11
		±0.22	±0.29	±0.40	±0.30	±0.24	±0.19	±0.14

Table: Bivariate correlation coefficient values for mercury accumulation by Rice

Days	*Variables*	*Correlations (r)*	*Significance level P <*	*Degrees of freedom*
15	Root	0.992	0.001	6
	Shoot	0.991	0.001	6
30	Root	0.998	0.001	6
	Shoot	0.996	0.001	6

11

Biotechnology and Marine Atmosphere

Marine Biotechnology

Humankind has always depended on the sea — for food, for medicinal treatments, for wealth. Now the new science of biotechnology has created new exciting possibilities for making the most of the ocean's bounty and for developing products never before associated with the sea. Biotechnologists use living organisms (or parts of organisms) to make or modify products, to improve plants or animals or to develop microorganisms for beneficial uses, including the development of materials that mimic molecular structures or functions of living organisms. Marine organisms, in particular, represent great phylogenetic diversity, making them reservoirs of unique genetic information and important natural resources for possible development. Exploratory research shows the great potential for exploiting the biochemical capabilities of marine organisms to provide models for new classes of pharmaceuticals, polymers, enzymes, other chemical products, and industrial processes, as well as vaccines, diagnostic and analytical reagents and genetically altered organisms for aquaculture and the seafood industry. Marine biotechology is also providing new tools and approaches for determining water quality and for understanding ecological relationships among marine organisms and for defining fisheries and stocks that will help improve marine resource management.

Marine biotechnology is an emerging field focused on investigating and developing technological applications of living marine organisms, their derivatives, and their bioprocesses. Products derived from marine organisms have been utilised to remediate the environment, increase food supply, develop new industrial processes and create new pharmaceuticals.

To date, examples of marine-derived pharmaceuticals include (1) an antibiotic from fungi; (2) two closely related compounds from a sponge that treat cancer and the herpes virus; and (3) a neurotoxin from a snail with painkiller properties 10,000 times more potent than morphine and without the side effects.

Biotechnology of Marine Extremophiles

"The sea is regarded as the origin of life on Earth and the oceans include the largest habitats on Earth hosting the most ancient forms of life. Over billions of years, marine microbes have moulded the global climate and structured the atmosphere. Knowledge of the biochemical processes that adapted, diversified and evolved in very different and extreme environments is the basis for discoveries in biotechnology". This statement reflects the current opinion of most of the scientific community in Europe. The fields of biotechnology that could benefit from mining the extremophiles are very broad and cover the search for new bioactive compounds for industrial, agricultural, environmental, pharmaceutical and medical uses. However, this potential remains to a large extent unexplored and, in respect of the drugs available on the market, only 30 per cent have been developed from natural products and so far less than 10 per cent have been isolated from marine organisms.

Extremophilic Archaea and Bacteria thrive in various extreme environments on our planet. They include thermophiles and hyperthermophiles colonising coastal hot springs, shallow and deep-sea hydrothermal vents, or psychrophiles thriving in deep ocean trenches, abyssal plains, polar oceanic regions and continental margin cold seeps. Extremophiles include also halophiles, thermoacidophiles and thermoalkaliphiles, named according to the properties of their respective environments. In addition, exploration of subterranean environments below the sea floor, through ODP, has revealed the existence of an important microbial biomass, most of it still uncultured and uncharacterised (Pedersen, 2000).

The potential impact of marine microbial resources on the biotechnology industry is mainly dependent upon biodiversity exploration and genomics dedicated programmes as well as specifically targeted biotechnological projects. Moreover, access to and preservation of the marine extreme environments is a major issue.

i. Besides microorganisms already available in public collections, exploration of extremophiles in the various extreme marine environments (hydrothermal vents, cold seeps, subterranean environments), either with cultural methods or total DNA molecular based approaches, is the most promising way to increase the probability of finding novel bioactive compounds.
ii. Access to sampling. The huge potential of marine extremophiles relies on access to sampling. The legal property is defined according to international regulations, and is therefore granted for EU member states in their EEZ. In international waters however, the property is in fact attributed to those able to sample the deep-sea, in fact a very small number of countries. Conservation of marine extreme environments has not been an issue up to very recently and this could change in the near future.
iii. Since the very first complete genome sequencing of

Haemophilus influenzae in 1995, genomic approaches have been successfully introduced in analyses of microbial genomes for gene identification and function assignment. Genomics and proteomics are therefore needed to insure the growth of databases, to discover novel genes and gene products through data mining. Expression libraries, constructed from total DNA extracted from crude samples, provide a short cut to gene repertoire without culture-based limitations.

iv. Specifically targeted biotechnological programmes are needed in order to characterise the properties of the novel bioactive compounds and assess their potential for development. When, as very often observed, the properties of the candidate molecule do not fit completely with the industrial requirements, engineering of the molecule is needed.

Exploration

Exploration of microbial biodiversity in these environments has opened a new era for microbiologists already blessed by the establishment of Archaea as a separate domain (Woese and Fox, 1977) and the successive introduction of molecular biology and genomics as basic and powerful tools. Besides the classic approach of strain isolation through culture, a new window has been opened during the last decade with the use of molecular markers to characterise the microbial biodiversity of various ecological niches (Amann *et al.*, 1996). One of the main conclusions is the growing emergence of clusters of "environmental clones" with no representative in culture, illustrating the "count-plate anomaly". The number of extremophilic species described has increased exponentially during the last two decades and encompasses Archaea, Bacteria and more recently some associated archaeal and bacterial viruses. Since the cultured strains already available could account for less than 1 per cent of the total number of species in marine extreme environments, *ecogenomics*, based on the complete sequencing of clones containing cosmids or bacterial artificial chromosomes with inserts from extremophiles, considerably extend the probability of finding new genes and their products suitable for the industry. Traditional and direct approaches to the finding of novel products are however still effective. For example, sampling of deep-sea hydrothermal vents microorganisms colonising selective substrates and subsequent cultures on corresponding selective media have led to the discovery of new thermostable esterases/lipases, proteases (keratinases) and xylanases.

Access to Samples and Marine Extremophiles

Access to hypersaline ponds and seas as well as shallow hot springs is usually easy and does not require heavy equipment. On the contrary, sampling of deep-sea continental margin cold seeps and hydrothermal vents is dependent upon specialised equipment such as vessels, man-operated submarines or ROVs. European countries possess all the basic equipment to fulfil that purpose, but increasing costs make integration necessary. Development of new samplers is also required in order to remain competitive with the US and Japan.

Genomics and Marine Microorganisms

Since microorganisms constitute the basis of any marine ecosystem whether they belong to photosynthetic, chemosynthetic or heterotrophic groups, their contribution to the functioning of marine ecosystems is of enormous importance. They are heavily involved in the recycling of major elements on a global scale. Knowledge of the genome sequences of key species is a prerequisite to study microbial responses to environmental changes. It is also the basis of gene mining for new potential biotech applications. Marine microbe complete genome projects account for approximately 15 per cent of the total genome projects and European contribution has been limited to one third of this work, the main effort being carried out by US consortia. This critically emphasises the need of a significant increase in means related to marine genomics in

Europe. Comparative genomics can be used to compare enzyme inventories of different extremophiles and complete already available biochemical data (hyperthermostable proteases for example (Ward *et al.*, 2002)) or to unravel discrepancies between experimental and genomic data, possibly suggesting new metabolic pathways. It significantly contributes to the identification of new genes and functions as demonstrated by the re-annotation of *Pyrococcus abyssi* (Cohen *et al.*, 2003). The rapidly expanding proteomic technologies are also expected to increase the list of proteins and enzymes involved in functional multiprotein complexes in DNA replication, transcription, translation and various metabolisms.

Microbial Marine Biotechnology: Brief View of Leading Fields

The Main Fields of Interest are Enzymes and Biopolymers: Enzymes from hyperthermophiles have been a major target of biotech programmes in the last decade (e.g. EU FP4 and FP5 "Extremophiles as cell factories"). Thermostable proteases, lipases, esterases, starch and xylan degrading enzymes have been actively sought and in many cases found in bacterial and archaeal hyperthermophilic marine microorganisms (Bertoldo and Antranikian, 2002). In some cases, the substrate specificity and the yield do not completely fit the industrial requirements and enzymes have to undergo an additional protein engineering step (e.g. DNA shuffling). The best known commercial success of thermostable enzymes is the Taq DNA polymerase, obtained from *Thermus aquaticus* (Yellowstone hot spring). Marine *Thermococcales* have been an important source of high fidelity thermostable DNA polymerases (Pfu, Vent, Pab, etc.) (Hamilton *et al.*, 2001), accounting for 30 per cent of total sales. In addition, the high structural conservation and complementation of DNA replication proteins between euryarchaeal *Pyrococcus* and human make hyperthermophilic archaea a model of choice to study eukaryotic DNA replication (Henneke *et al.*, 2000).

Microbial exopolysaccharides (EPS), especially those produced by some mesophilic Vibrios and *Alteromonas* strains isolated from deep-sea hydrothermal vents, display interesting properties currently under evaluation for therapeutic uses. The main fields concerned are tissue regeneration and cardiovascular diseases (antithrombotic/proangiogenic effects). Characterisation and preliminary results on anti-coagulant activities of the EPS showed that native EPS were deprived of effects whereas sulphated derivatives were active (Guezennec, 2002).

Characterisation of Sulphate Assimilation in Marine Algae Focusing on the Enzyme 52 -Adenylylsulphate Reductase 52-Adenylylsulphate (APS) reductase was characterised in diverse marine algae. A cDNA encoding APS reductase from *Enteromorpha intestinalis* (EAPR) was cloned by functional complementation of an *Escherichia coli cysH* mutant. The deduced amino acid sequence shows high homology with APS reductase (APR) from flowering plants. Based on the probable transit peptide cleavage site the mature protein is 45.7 kD. EAPR expressed as a His-tagged recombinant protein catalyses reduced glutathione-dependent reduction of APS to sulphite, exhibiting a specific activity of approximately 40 μmol min^{-1} mg $protein^{-1}$ and Michealis-Menten kinetic constants of approximately 1.4 mm for reduced glutathione and approximately 6.5 μm for APS. APR activity and expression were studied in relation to the production of 3-dimethylsulphoniopropionate (DMSP), a sulphonium compound produced by many marine algae.

A diverse group of DMSP-producing species showed extremely high enzyme activity (up to 400 times that found in flowering plants). Antibodies raised against a conserved peptide of APR strongly cross-reacted with a protein of 45 kD in several chlorophytes but insignificantly with chromophytes. In the chlorophyte *Tetraselmis* sp., APR activity varies significantly during the culture cycle and does not follow the changes in cellular DMSP content. However, a positive correlation was found between cell-based APR activity and specific growth rate.

S is an essential nutritional element for plant growth and development. Available in the environment primarily as sulphate it is imported into cells where it is incorporated into organic molecules as sulphate (e.g. extracellular polysaccharides like agar) or is reduced to sulphide before being incorporated as the thiol group of Cys. Cys serves as the central intermediate from which a diversity of primary and secondary metabolites and coenzymes are derived (Leustek and Saito, 1999). S-containing metabolites participate in a variety of cellular processes, including disease resistance (Alsher, 1989; Booth and Walker, 1997), tolerance to oxidation (Lappartient and Touraine, 1997), heavy metals (Ruegsegger and Brunold, 1992; Schafer *et al.*, 1997), water stress (Storey *et al.*, 1993), and developmental signalling (Bouarab *et al.*, 1999).

Although S research in vascular plants has a 30-year history, only recently has the process been studied at the molecular level. Molecular and biochemical evidence has demonstrated significant differences in the first enzyme-catalysing sulphate reduction among vascular plants, bacteria, and fungi. In all of the organisms, sulphate assimilation begins with the enzyme ATP sulphurylase that catalyses the adenylation of sulphate to 52-adenylylsulphate (APS). Adenylation is necessary because sulphate is relatively unreactive. APS is then reduced by APS reductase in plants and some bacteria (Bick and Leustek, 1997; Abola *et al.*, 1999; Bick *et al.*, 2000), whereas in other bacteria and fungi APS is further phosphorylated at the 32 position forming 32-phospho-52-adenylylsulphate (PAPS) before being reduced by PAPS reductase. Plant APR is unique in that it is able to use reduced glutathione (GSH) as a source of electrons. By contrast, bacterial APR and PAPS reductase require the redox factors thioredoxin (Trx) or glutaredoxin (Grx) as a source of electrons. The GSH-dependency of plant APR is probably mediated through a carboxyl terminal domain that functions as a Grx, which is lacking in the bacterial and fungal enzymes. In flowering plants, APR is thought to be a primary regulation point for the sulphate assimilation pathway (Leustek and Saito, 1999).

Most of what is known about sulphate assimilation in plants is from research on flowering plants. Little is known about the process in algae. Marine algae metabolise S in unique ways, thus they likely have special S requirements. For example, many produce large amounts of sulphated extracellular polysaccharides like agar. Others produce and accumulate in abundance a sulphonium compound known as 3-dimethylsulphoniopropionate (DMSP). DMSP synthesis is limited to very few species of flowering plants (Hanson and Gage, 1996) but is widespread among marine algae (Keller *et al.*, 1989; Grone, 1995).

It probably serves as a compatible intracellular osmolyte and can accumulate to very high concentrations. Up to 0.5 m has been reported for some species (Keller *et al.*, 1989). DMSP is synthesised from Met (Gage *et al.*, 1997), therefore, it is dependent ultimately on the activity of the sulphate assimilation pathway. There is evidence that in some species DMSP synthesis is stimulated under N-limiting conditions (Grone and Kirst, 1992). DMSP is the primary source for one of the most important biogenic gases in the atmosphere, dimethylsulphide (DMS), accounting for nearly 50 per cent of the global biogenic S input into the atmosphere (Andreae and Raemdonck, 1983; Malin *et al.*, 1992; Grone, 1995).

As the first step towards understanding S metabolism in marine algae APR was studied in a variety of species, and a cDNA encoding the enzyme was cloned from the marine chlorophyte *Enteromorpha intestinalis*. The cDNA clone was used to express the enzyme for study and for production of a specific antibody. This antibody was used to study the regulation of APR in relation to DMSP production in another chlorophytic species *Tetraselmis* sp.

Cloning and Analysis of an APR cDNA from E. Intestinalis

Functional complementation of an *Escherichia coli cysH* mutant with a cDNA library prepared from *E. intestinalis* resulted in the isolation of approximately 100 positive clones after screening 1×10^6 independent transformants. Twenty-five clones were selected for sequence analysis. All were found to be derived from the same gene but varied in the length of their 52 and 32 ends. The longest clone, referred to as EAPR, was 1,525 bp. It shows a continuous open reading frame beginning from nucleotide 1 to the termination codon at 1,336 to 1,338. Therefore, the translation initiator is not possible to predict, although a Met codon exists at position 68 to 70. An alignment of the EAPR open reading frame with Arabidopsis APR3 shows that the two have a high level of homology (56 per cent identity, 63 per cent similarity). The first 41 amino acids of EAPR show the properties of a transit peptide for plastid localisation with a possible cleavage site at position 32 to 35 (LRAG) (Von Heijne *et al.*, 1989). APR is known to be plastid localised in flowering plants (Rotte, 1998). If the EAPR protein were processed at the putative cleavage site (LRAG) the mature protein would be 411 amino acids with a predicted mass of 45.7 kD.

The alignment shows that EAPR contains all of the hallmarks of plant APRs. The region from amino acids 83 to 310 is homologous with the sulphate assimilatory APR and PAPS reductase of bacteria and fungi (Berendt *et al.*, 1995; Bick *et al.*, 2000). This region has been defined as the reductase domain of plant APR (Bick *et al.*, 1998) and includes a PP motif (indicated by dots), likely involved in nucleotide binding, and a region conserved in all of the assimilatory (P)APRs (indicated by asterisks) that is important for catalytic activity (Berendt *et al.*, 1995). The region from amino acids 311 to 445 shows homology with Trx and Grx including the active site sequence (indicated with diamonds).

The genomic organisation of the EAPR gene was studied in *E. intestinalis* by Southern blotting using the EAPR cDNA as a probe. A simple hybridisation pattern was observed with all of the restriction enzymes tested. This indicates that the sequence homologous to the EAPR cDNA likely exists as a single locus in the alga.

Characterisation of the EAPR Product

E. coli JM96 strain (*cysH* mutant) is devoid of the enzyme responsible for the reduction of sulphate to sulphite (PAPS reductase). However, the strain expressing EAPR showed an APR activity of approximately 2×10^{-3} units mg^{-1} in the crude bacterial extract. Comparable APR activity was observed with a deletion subclone of the EAPR cDNA from which 249 bp was removed from the 52 end, indicating that the catalytic core of the enzyme lies beyond the 84th amino acid.

Plant APRs are unable to function with electrons donated from sources other than thiol compounds despite the existence of a C-terminal domain with sequence homology to Trx. To determine what electron donor EAPR protein uses during complementation, a test was performed using *E. coli* strains with mutations in gshA (a-glutamyl-Cys synthetase) or gshB (GSH synthetase) required for GSH synthesis, or trxB (Trx reductase). The result indicates that EAPR cDNA clone was able to complement a *cysH* mutant of *E. coli* that also carries the trxB mutation. However, it was unable to complement the strains lacking GSH (*E. coli* strains carrying *cysH* and gshA or gshB), although active EAPR protein was expressed in these bacterial strains as demonstrated by the in vitro APR activity measurements. Consequently, it is concluded that EAPR protein uses GSH, rather than Trx reductase, for catalysis.

The purified recombinant EAPR product has no PAPS reductase activity and exhibited a specific APR activity of approximately 40 units mg^{-1}. This activity is significantly greater than the reported activities of APR products from Arabidopsis (0.2-4 units mg^{-1}; Bick *et al.*, 1998) but similar to APS reductase from Lemna (Suter *et al.*, 2000). Moreover, the EAPR protein is not sensitive to inactivation by freeze thawing and storage in the thawed state as are Arabidopsis APRs, suggesting that the algal enzyme is a much more active and robust form of APR. The kinetics of the algal APR, however, is comparable with that reported for Arabidopsis APRs with apparent K_ms of 1.39 mm for GSH and 6.5 ìm for APS.

A strong temperature dependence of the activity was observed for the purified recombinant APR. A sharp break was found between 20°C to 25°C with an apparent E_a, estimated from the slope of the Arrhenius plot, of 10.1 kcal mol^{-1} for the temperatures between 0°C to 20°C. The Q_{10} below 20°C was estimated to be 1.9, comparable with those found for many other biological processes (Berry and Raison, 1981). Similar responses were observed for APR activity in cell-free extracts of the phytoplankton *Tetraselmis* sp. and *Emiliania huxleyi*.

Enzyme Activity and Antigenicity of APR from E. Intestinalis and Several Marine Phytoplankton

APR and PAPS reductase activity were determined in several major groups of marine algae, including three chlorophytes (*E. intestinalis*, *Tetraselmis* sp., and *Dunaliella salina*), two diatoms (*Thalassiosira weissflogii* and *Thalassiosira oceanica*), two prymnesiophytes (*I. galbana* and *E. huxleyi*), and one dinoflagellate (*Heterocapsa triquetra*). These algae are from diverse phylogenetic origins and have different capacities for DMSP production. The results showed that most of the species tested have high APR activities ranging from approximately 100 up to 500 × 10^{-3} units mg^{-1}, except for the dinoflagellate species. These activities are 100-fold greater than that from young leaves of two species of flowering plants Arabidopsis and Indian mustard (approximately 1 × 10^{-3} units mg^{-1}). No significant PAPS reductase activity (< 1 per cent of that for APR activity) was detected in any of the species examined, indicating that APR is likely the primary route for sulphate reduction in these organisms.

The cross-reactivity of an APR-specific peptide antiserum was tested against cell lysates from the algal species. The antibody cross-reacted significantly only with the chlorophytic species. The cross-reacting peptide in these species was approximately 45 kD, a size that is consistent with the mass of the mature protein predicted from the EAPR cDNA clone. The strongest cross reactivity was found with *Tetraselmis* sp., correlating with the extraordinarily high APR activity found in this species, suggesting that this high activity is achieved in part by high expression level of APR in this organism. Weak, specific cross-reactivity was observed with the prymnesiophyte, diatom, and dinoflagellate species examined, suggesting that APR from these species are divergent in the peptide sequence region (IAFSGAEDVA) conserved among APR from flowering plants and green algae. Similar species divergence in enzyme antigenicity has been observed for enzymes involved in other assimilatory pathways in marine algae, such as nitrate reductase and Gln synthetase (Gao *et al.*, 1993; Robertson *et al.*, 1996). The current result also supports the idea of a diverse evolutionary history for marine organisms.

Regulation of APR Activity and DMSP Production in Tetraselmis sp.

The high APR activity found in most marine algae examined in this study may reflect the special S needs of these organisms. Most of the species examined are known to produce large amounts of DMSP when N is limiting in the environment (Grone, 1995). DMSP synthesis may ultimately be dependent on the activity of APR as it was demonstrated earlier that this is the first enzyme involved in sulphate reduction in these organisms and is most likely the regulation point in the sulphate assimilation pathway of flowering plants (Brunold, 1981; Brunold and Suter, 1984). As a first step towards understanding environmental regulation of APR and its role in the production of DMSP in marine organisms APR activity and intracellular DMSP contents were examined in *Tetraselmis* sp. cultures grown with limiting or sufficient N supplies. These conditions are known to alter the organism's ability to synthesise DMSP.

For both of the treatments the highest activities in *Tetraselmis* sp. cells observed during the early log growth phase, the most rapid growth stage of batch cultures. As the cultures became depleted in N (d 3–5), APR activity decreased to <40 per cent and reached a 10 per cent to 25 per cent level when the cultures reached stationary phase. For the abundant N treatment the pattern for cell-based APR activity was similar to that based on cellular protein content in these cultures. In contrast, APR activities in the cells grown with limited N supply, when expressed on cellular protein basis, did not change significantly during the culture cycle, although slightly higher activities were observed during the mid-log growth phase of these cells. The APR activities were comparable with those found in the early log growth-phased cells with abundant N supply. When the APR activity was expressed on a per-cell basis, the pattern of the activity change was similar to that found for cultures grown with abundant N supply with maximal activity during the rapid growth stage and low activity after N was depleted from the cultures. This result may reflect the organisms' ability to quickly mobilise cellular proteins after encountering N shortage. Since APR activity on protein basis was maintained through out the culture cycle, the results suggest that it may be spared from deactivation or proteolysis. Immunoblot analysis revealed that the changes in APR activity observed here are paralleled by changes in APR protein abundance.

It is interesting that under both of the growth conditions, APR activity on cell basis began declining when medium nitrate was reduced to approximately 35 μm and reached a plateau 1 d after the depletion of N from the media, suggesting a coordinated regulation between sulphate and nitrate assimilation. Moreover, the cell-based APR activity in cells grown with limited N supply plateaued, upon the depletion of nitrate from media, at a level twice that in cells grown with abundant N supply.

Given the observation that cell-based APR activity is inversely correlated with the growth curve and N availability under both of the growth conditions, a close coupling of grow rate and cell-based APR activity was hypothesised. Comparison of specific growth rate (μ) with cell-based APR activity further support the hypothesis. The result suggests that cell-based APR activity can be used as an index for estimating growth rate and thus productivity as well as the N-nutrition state of the cells.

Intracellular DMSP content in *Tetraselmis* sp. cells grown under limited or abundant N supply conditions. Both of the cultures exhibited similar intracellular DMSP content (approximately 15 fmol cell^{-1}) when N was present in the medium. However, the increase in DMSP content after N was depleted was much more rapid in cells grown with limited N supply than those grown with abundant N supply. Further, after reaching stationary phase, the N-limited cultures attained a higher intracellular DMSP content compared with cells grown with abundant N. The changes in cellular DMSP content did not correlate with the changes in APR activity, suggesting that APR does not directly limit or regulate DMSP synthesis and that the synthesis of DMSP may be partly or entirely supported by intracellular resources of assimilated S.

S metabolism is important to marine algae in that they produce and accumulate to high concentrations S compounds that most terrestrial plants do not produce. Unlike most terrestrial crop plants, many marine algae live in habitats that are characterised by limited N and abundant S supply. The concentration of sulphate in seawater is approximately 40 mm (Pickard and Emery, 1988). In response to this environment, many have evolved the ability to produce DMSP, a functional analog of the N-containing osmolyte Gly betaine. In those species that accumulate it, DMSP is likely a major sink for assimilated S in addition to protein, especially when N is not available. The ability to synthesise DMSP may play a critical role in the survival of these organisms in their environmental niches.

In the present study, as the first attempt to understand sulphate assimilation in marine organisms, a cDNA (EAPR) from *E. intestinalis* was cloned. The result indicates the structural and functional similarities between algal APR from marine chlorophyte algae and flowering plants. In addition, the result demonstrated the use of functional complementation approach as a powerful tool in investigating molecular and biochemical features of processes in marine algae.

Examination of temperature response of APR activity showed that the enzyme activity is strongly temperature dependent with a maximal activity at 20°C to 25°C. A sharp break at low temperatures (<30°C) for enzyme activity has only been observed for nitrate reductase previously, the first enzyme involved in nitrate reduction (Gao *et al.*, 1993), but not for other enzymes that are also important in assimilatory processes in plants, e.g. RuBP carboxylase, PEPC Kinase, Gln sythetase, and Glu dehydrogenase. However, unlike APR where a similar pattern was seen in both of the chlorophyte and chromophyte, a sharp break at low temperatures (15°C) for nitrate reductase was only observed in chromophytes. The chlorophytic nitrate reductase activity was linear up to 35°C (Gao *et al.*, 2000). Such large differences in temperature dependence for important assimilatory enzymes may have significant impact on the geographic distribution and seasonal abundance of these organisms in the marine environment.

A significant correlation between the specific growth rate and APR activity on a per cell basis was found in *Tetraselmis* sp. cultures. Together with the observation that APR activity declines as N is depleted from the medium in cells grown with both of the limiting and the abundant N supply, the results suggest a coordinated regulation between sulphate and nitrate assimilation. These interactions were thought to reflect a mechanism to coordinate and balance the flow of the two essential elements (S and N) into protein (Reuveny *et al.*, 1980; Brunold and Suter, 1984). The previous studies showed that low nitrate concentration causes a drastic decrease in the specific APR activity after 24 h (Brunold and Suter, 1984) as well as the influx of sulphate into cells (Clarkson *et al.*, 1989) in flowering plants. The reduction in APR is due to a decrease in the steady-state level of mRNA and protein (Lee, 1998). These results are consistent with our finding that high level of APR activity in the marine alga *Tetraselmis* sp. was observed when nitrate is plentiful. It is interesting that cell-based APS activity in cells grown with limited N supply plateaued at a level twice that found in cells grown with abundant N supply. One possible explanation for this observation is that APR is spared from proteolysis relative to other proteins in these cells.

The experiments did not reveal a direct correlation between APR activity and DMSP production as indicated by the intracellular DMSP content. This may be explained in part by the possibility that intracellular DMSP content does not represent accurately the rate of DMSP production. It also, in part, suggests an ability of the alga to mobilise intracellular S resources under the stress of N shortage. In this regard, the extraordinarily high APR activity, compared with flowering plants, may be of significance. Thus, whereas APR does not directly limit or regulate DMSP synthesis, its high activity may reflect a process for building an intracellular pool of reduced S, which could serve as a resource for support of DMSP biosynthesis when needed. To obtain a comprehensive understanding of the role of APR in DMSP production and its regulation in the marine organisms that have special S needs, further studies are required.

Algal Materials and Growth Conditions

Enteromorpha intestinalis and *Ulva pertusa* were collected in early summer at the intertidal zone of the Gateway National Recreation Area at Sandy Hook, New Jersey (US Department of the Interior collection permit no. SHU–98–03). The tissue was rinsed with ultrapure water and frozen immediately with dry ice. Frozen samples were stored at "80°C until use. Phytoplankton species (*Tetraselmis* species, CCMP897; *Dunaliella salina*, ATCC30861; *Isochrysis galbana*, CCMP1323; *Emiliania huxleyi*, CCMP1299, *Thalassiosira weissflogii* CCMP1336, *Thalassiosira oceanica* CCMP1005, and *Heterocapsa triquetra* CCMP449) were grown in f/2-enriched seawater medium (Guillard, 1975) at 19°C on a 12-h:12-h light:dark cycle at 200 ìmol m^{-1} s^{-1}, using NO_3^- as the sole N source. Artificial seawater medium (Sigma S-9883) with f/2-enrichments was used in the experiments assessing the role of APR in relation to DMSP production. The species with CCMP designations were obtained from the Provasoli-Guillard National Centre for Culture of

Marine Phytoplankton (Bigelow Laboratory for Ocean Sciences, West Boothbay Harbor, ME). *D. salina* was obtained from the American Type Culture Collection (Fairfax, VA). All of the samples for analysis were harvested at noon by centrifugation, frozen in liquid N_2, and stored at -80°C.

Construction of a cDNA Library from *E. intestinalis* and Cloning of EAPR Total RNA was prepared as follows. Twenty-three grams (fresh weight) of frozen *E. intestinalis* was ground into powder with liquid N_2 then mixed with 60 mL of extraction buffer containing 4 m guanidine thiocyanate, 25 mm sodium citrate (pH 7.0), 0.5 per cent (w/v) sarcosyl, and 0.7 per cent (v/v) a-mercaptoethanol. Then 0.1 volume of 2 m sodium acetate (pH 5.0) was added. The mixture was extracted with 40 mL of phenol (diethyl pyrocarbonate-water saturated) followed by an addition of 8 mL of chloroform:isoamyl alcohol (49:1). The mixture was centrifuged at 5,000*g* for 20 min at 4°C. The upper aqueous phase was transferred into a clean tube and re-extracted with 40 mL of chloroform:isoamyl alcohol (49:1). Total RNA was precipitated with an equal volume of ice-cold isopropanol. The RNA pellet was dissolved in 20 mL of extraction buffer and reprecipitated with 2 volumes of ethanol. The cDNA library was constructed in λ*TriplEx* using the *Eco*RI and *Xba*I cloning sites. The library contained 1.2×10^6 independent clones.

The λ-phage cDNA library from *E. intestinalis* was converted to plasmid form by mass excission according to the manufacturer's instruction (CLONTECH Laboratories, Palo Alto, CA). The resulting plasmid library was then used to screen for clones that complement the *Cys* requirement of *Escherichia coli* strain JM96 (*cysH*). JM96 was provided by the Coli Genetic Stock Centre (Yale University, New Haven, CT).

DNA Sequencing and Nucleic Acid-Blot Analysis

DNA sequencing was carried out on both of the strands using plasmid primer sites with the original EAPR cDNA clone as well as deletion subclones on the parent vectors, using a DNA sequencer.

For Southern blotting, *E. intestinalis* genomic DNA was isolated from materials obtained. Four grams of tissue (frash weight) was first pulverised in liquid N_2 and then mixed with 23 mL of preheated (65°C) extraction buffer containing 100 mm Tris [Tris(hydroxymethyl)-aminomethane]-HCl (pH 8.0), 500 mm NaCl, 8.3 mm NaOH, 1.25 per cent (w/v) SDS, and 0.38 per cent (w/v) sodium bisulphite. The mixture was then incubated at 65°C for 15 min, and 7 mL of 5 m potassium acetate was added. After standing on ice for 20 min, the mixture was centrifuged at 4,000*g* for 10 min. The DNA in the supernatant was precipitated by adding 0.7 volume of prechilled isopropanol and washed with 70 per cent (v/v) ethanol. The DNA pellet was then resuspended in T5E buffer (50 mm Tris-HCl, pH 8.0, and 10 mm EDTA), containing DNase-free RNase A (final concentration of 20 ìg/mL) and incubated at 37°C for approximately 1 h. The DNA was then subjected to buffer-saturated phenol/chloroform extraction twice and reprecipitated with ethanol. The final DNA pellet was resuspended in TE buffer (10 mm Tris HCl, pH 8.0, and 1 mm EDTA) and used for Southern blotting.

After digestion with the appropriate restriction enzymes, the genomic DNA were electrophoresed in 0.8 per cent (v/v) agarose gels and transferred onto Zeta-Probe membrane (Bio-Rad Laboratories, Hercules, CA). The DNA probe, labelled by the random primer method (Life Technologies/Gibco-BRL, Cleveland) using [α -^{32}P]dCTP, was made from the full-length cDNA of EAPR1. Hybridisation was performed in 2× SSC containing 7 per cent (w/v) SDS at 50°C.

Construction of Expression Plasmids and Purification of Recombinant EAPR

The EAPR cDNA, after removing 189 nucleotides from the 52 end, was sucloned as a 1,360-bp *Not*I fragment into pET-30b (Novagen, Madison, WI). The plasmid, pET-EAPR was transformed into BL21(DE3) plysS. Transformants were grown at 37°C to an optical density of approximately 0.4 at 600 nm and then expression was induced with 1 mm isopropyl-a-dthiogalactoside for 2 h before harvesting. Because of the association of a 55-amino acid-long fragment (which contains His-tag, thrombin, S-tag, and enterokinase cleavage site) with the recombinant EAPR enzyme, it is estimated that the resulting polypeptide is 466 amino acids with a predicted mass of approximately 52 kD.

The purification of recombinant EAPR enzyme was conducted using Ni-agarose (TALON, CLONTECH Laboratories). All of the procedures were carried out at room temperature (22°C–25°C). The bacterial cell pellet from a 100-mL culture was lysed by sonication in 10 mL of 20 mm Tris-HCl, pH 8.0, and 100 mm NaCl. The lysate was centrifuged at 12,000*g* for 10 min, and the supernatant was incubated with 1.5 mL of TALON resin for 30 min on a rotating shaker. The resin was washed with 5 volumes of lysis buffer and then packed into a disposable column. The column was further washed with 5 volumes of lysis buffer containing 10 mm imidazole. EAPR protein was eluted with lysis buffer containing 50 mm imidazole, then the buffer was replaced with 100 mm Tris-HCl, pH 8.5 containing 1 mm EDTA using microconcentrator unit (Microcon-30, Millipore, Bedford, MA). The pure protein was stored at -80°C.

The EAPR cDNA was also subcloned as a 1,310-bp (*Sst*I and *Hind*III) fragment into pET-30a. The plasmid was transformed into BL21(DE3) plysS and induced only for APS activity determination.

(P)APS Reductase Activity and Protein Assays

Algal cells were harvested, frozen in liquid N_2, and stored in -70°C before analysis. Algal cells were lysed in 100 mm Tris-HCl, pH 8.5, 1 mm EDTA, and the supernatant collected after centrifugation at 14,000*g* for 15 min. Microalgae were lysed by sonication. Macroalgae were ground to a fine powder in liquid N_2. Enzyme activities were measured as described by Bick *et al.* (1998) using GSH (1 mm) as the electron donor and APS (approximately 500 Bq $nmol^{-1}$) as subtrate. One unit of enzyme activity is defined as micromoles of product formed per minute. Apparent K_m values were calculated by direct-fit to the Michaelis-Menten equation using Sigma-Plot (Jandel Scientific, version 5.0, SPSS, Inc., Chicago). Protein was estimated by Bradford assay (Bradford, 1979) using a bovine serum albumin standard.

Antibody Preparation and Immunoblotting

Antibodies were prepared in rabbits by Biosynthesis (Lewisville, TX) using the MAP peptide method. A peptide sequence conserved among APRs from flowering plants and EAPR was chosen for the procedure.

Immunoblotting was carried out as described by Wang *et al.* (1993). The antibody was used at a dilution of 1:5,000. The secondary antibody was horseradish peroxidase-linked goat anti-rabbit diluted 1:10,000. The immune complexes were detected with the Renaissance Kit (DuPont NEN, Inc., Boston) and quantified with a densitometric video camera (model: Datavison 261 from Fernon Electronic Imaging, St. Michaels, MD) and the software NIH Image (version 1.0) provided by the National Institutes of Health (Bethesda, MD).

DMSP Measurements

DMSP content was measured indirectly as DMS produced after alkaline treatment of algal cells (White, 1982). Algal cells were collected by centrifugation and resuspended in 100 mm Tris-HCl, pH 8.0. Five to 10 μL of the cell suspensions was placed onto filter papers (8-mm diameter disc). The filter papers were placed into 2-mL glass vials containing 0.4 mL of 25 per cent (w/v) NaOH and capped with Teflon-coated septum. The vials were incubated at room temperature for 4 h and then placed at 4°C overnight. DMS produced in the head space of the vials was removed with a gas-tight syringe and measured by gas chromatography on a Poraplot Q, fused silica column (10 × 0.53 mm, Chrompack, The Netherlands) using a Shimadzu GC-17A gas chromatograph outfitted with a flame ionisation detector.

Bacteria Pick Up Genes for Antibiotic Resistance

Scientists have confirmed their suspicions that a strain of hospital bacteria *Staphylococcus aureus* acquired resistance to the powerful drug vancomycin from another species of bacteria also commonly found in hospitals.

The two bacteria were present in the wound of a patient in a Michigan hospital in June 2002.

The confirmation came when researchers sequenced DNA from *S. aureus* and concluded that the DNA, which carried genes for vancomycin resistance, had almost certainly come from the bacterium *Enterococcus faecalis*. Apparently the resistance genes had "jumped" from one microbe to another.

Scientists have known that in the laboratory the two species can exchange genes for resistance to vancomycin. But the new study marks the first time scientists have documented how the exchange occurred in real life.

Scientists at the Centres for Disease Control and Prevention in Atlanta, Georgia, led the research. The findings appear in *Science*.

"The really important observation is that this [exchange] happened in a clinical environment," says Steven R. Gill of The Institute for Genomic Research (TIGR) in Rockville, Maryland, where the DNA was sequenced.

S. aureus is already resistant to most antibiotics, and scientists had long-anticipated that the bacterium would become resistant to vancomycin, leaving physicians without a last-resort treatment. More than a year ago, doctors' fears came true when they discovered *S. aureus* resistant to vancomycin in a Michigan dialysis patient. Staph bacteria can still be treated with some antibiotics.

Bacterial genomes have large amounts of DNA that can move around on chromosomes, between organisms, and even jump between species. This enables them to pass traits such as drug resistance from one to another.

Antibiotic resistance is a growing health concern, which led the US Department of Health and Human Services to launch a campaign in September called "Get Smart: Know When Antibiotics Work" to alert doctors and patients to be cautious in using antibiotics.

Mobile DNA: Genomic Studies Illuminate Antibiotic Resistance

In September 2002, a patient with a chronic foot ulcer checked into a Pennsylvania hospital. Doctors later discovered that the ulcer contained *Staphylococcus aureus* bacteria, one of the most common causes of hospital infections.

The bacteria were resistant to vancomycin, the antibiotic of last-resort in fighting infections. It was only the second case of vancomycin-resistant staph reported in the United States. The Pennsylvania physicians also had evidence the bacteria had acquired these resistance genes from another species of bacteria often found in hospitals, *Enterococcus faecalis*.

Antibiotic resistance is a frightening reality. Now, scientists are a step closer to understanding how antibiotic resistance spreads in the microbial world because the genomes of two gut bacteria have been sequenced.

The microbes, which usually live harmlessly in our intestines, are *Bacteriodes thetaiotaomicron* and *Enterococcus faecalis*.

Their genomes contain an incredible amount of mobile elements—DNA that can move around on chromosomes, among organisms and even between species. Mobile DNA can carry genes for virulence and drug resistance, as well as benign genes.

"These mobile elements give bacteria the ability to quickly pass on traits," says Ian T. Paulsen, who led the enterococcus genome-sequencing project at the Institute for Genomic Research (TIGR) in Rockville, Maryland.

Until sequencing the entire genome, scientists did not realise that a quarter of enterococcus' genome is made-up of mobile

DNA. In fact, its genome contains one of the highest percentages of mobile elements ever seen in bacteria. Within these regions are genes for vancomycin resistance and for virulence.

It had long been established that enterococcus has mobile elements, says Gary M. Dunny of the University of Minnesota in Minneapolis, who studies the bacterium.

"But until now, we have never had a picture of the entire genome of one strain and the total number of mobile elements, which is remarkably high," he says. TIGR sequenced a strain from a patient with the first case of vancomycin-resistant *E. faecalis* in the United States.

Bacteriodes is also rich in mobile elements, although these do not harbor antibiotic resistance genes. Jeffrey I. Gordon of Washington University School of Medicine in St. Louis, Missouri, led the bacteriodes project.

"Bacteriodes is not carrying these genes now, but they can pick them up," says Abigail A. Salyers of the University of Illinois in Urbana-Champaign. Salyers has also served as President of the American Society of Microbiology.

"There is much greater potential for transferring resistance genes than we thought," she adds.

The genomes of bacteriodes and enterococcus reinforce something scientists have been concerned about for years: the remarkable fluidity of the bacterial gene pool. This fluidity allows bacteria to exchange DNA to enhance their ability to cause disease or their resistance to antibiotics.

"It's really becoming a problem now because we're running out of antibiotics," says Michael S. Gilmore of the University of Oklahoma in Oklahoma City. Gilmore studies enterococcus, and wrote a commentary accompanying the two papers in *Science*.

Bacteriodes, enterococcus and staph are just a few examples of bugs that are becoming increasingly difficult to treat. The widespread use of antibiotics puts selective pressure on only the hardiest bacteria to survive—which often carry virulence or drug resistance genes.

"We're facing something we've never faced before—the loss of a cure," says Salyers. New antibiotics are expensive and difficult to develop and "pharmaceutical companies are shutting down antibiotic programmes," she adds.

Hospitals are breeding grounds for some of these dangerous microbes. When you enter a hospital, your intestines carry a normal consortium of microbes. But taking antibiotics can kill susceptible bacteria and leave room for resistant enterococci to take up residence.

These bacteria are lurking on medical instruments and surfaces—just waiting to be ingested and find a niche in your body. The drug-resistant, opportunist enterococci are then poised to infect other parts of the body like surgery wounds, the urinary tract and bloodstream.

The gut environment also contributes to the spread of dangerous genes because enterococcus and bacteriodes live along with over 500 species of bacteria. In fact, there are more bacteria in our intestine than cells in our body.

These gut bacteria are in contact with one another and with other bacteria that pass through the intestine, where they can swap virulence and resistance genes.

"Your colon is a like a singles bar," says Salyers. "Bacteria are passing DNA around like there's no tomorrow."

These bacteria don't just stay in your body, but pass into the water, soil and food supply we all share.

Mutant Bacteria and the Failure of Antibiotics

Hospitals have germs. And germs have a remarkable ability to develop resistance to the antibiotics we rely on to kill them. These facts are well known to the medical community and have been a source of great concern for years. A number of books and articles for both the lay reader and the professional have been written on the subject, each in its way raising a red flag that says, in one way on another, we have to control the use of antibiotics less we lose them to smart bacteria that learn to become immune to their toxic effects.

Michael Shnayerson and Mark J. Plotkin, authors of The Killers Within: The Deadly Rise of Drug-Resistant Bacteria, have now joined the army of alert writers who are concerned about the dangers of antibiotic resistance and its implications for the return of infectious diseases that cannot be effectively treated. The authors argue that today's hospitals are not only spreading bacteria from patient to patient, but are also harbouring a tougher breed of bug—one that is resistant to antibiotics.

Most infections are either bacterial or viral, and bacterial infections are susceptible to antibiotics. These drugs are designed to attach to enzymes on bacterial cell walls, either preventing the microbes from replicating or killing them outright. Unless, that is, the bacteria mutate and change their enzymes, thus preventing the drug from attaching.

Penicillin was greeted as a panacea when it was developed in the early-20th century. In the decades since, the authors claim—as have others before them—that the overuse of antibiotics for every possible illness has "educated" bacteria, creating opportunities for mutations to occur: "If misuse of antibiotics created drug resistance in the first place, poor infection control in hospitals allowed the bugs to spread."

Bacteria may be small, but these one-celled organisms can divide and reproduce into more daughter cells than the human population of Earth in just fourteen hours.

Plotkin, an ethnobiologist, and Shnayerson, a contributing editor to *Vanity Fair*, report what they have learned from interviews and the scientific literature about how bacteria develop resistance and the role genes play in this process. They express admiration for bacteria's clever defence mechanisms, including the evolution of enzymes that attack antibiotics and tiny pumps that vomit the drug out of the cell.

In the first half of this book, the authors try to scare the general public into recognising the seriousness of the threat. Yet despite the scare tactics, the authors succeed in creating a suspenseful narrative.

Indeed, their claim that the medical industry has not taken growing drug-resistance seriously enough and has failed to improve antibiotics suggests a fatal conclusion fitting of Stephen King: People are dying of bacterial infections that were treatable a few years ago.

By shadowing scientists around the globe—including the "genetic detectives" and "microbe hunters" who work with the world's most dangerous pathogens, as well as epidemiologists investigating outbreaks of bacterial infection—the authors craft an informative thriller with vivid descriptions and tales of scientific sleuthing.

The book tells the story, for instance, of William Noble, a microbiologist at St. John's Hospital for Diseases of the Skin in London, who, in the early 1990s, created a strain of *Staphylococcus aureus* that was resistant to the antibiotic vancomycin by exposing the microbe to another bacterium with resistant genes. A few years later, a vancomycin-resistant *S. aureus* was isolated from a lung-cancer patient in Japan, suggesting that Noble's laboratory experiment had happened in nature.

Like hospitals, the meat industry is a source of rising drug-resistance. Small doses of antibiotics (called "growth-promoters") added to animal feed contribute to the development of resistant strains by familiarising bacteria with the drugs without actually threatening them. The authors argue that substantial research shows that these resistant bacteria are easily transferred to humans eating these animals.

Towards the end of the book, the authors leave behind the horrors of resistant bacteria and begin to describe the search for new antibiotics. Until recently, most natural antibiotics have been found in soil and fungi. Among the scientists trying novel approaches to discovering antibiotics are researchers who collect saliva from lizards in Indonesia and distill sewage water in the former Soviet Republic of Georgia.

These researchers are experimenting with animal peptides and miniscule viruses that act as natural antibiotics. Peptides punch their way through the bacterial cell membrane regardless of these enzymes.

Scientists in Georgia have been using viruses called "phages" for decades to puncture the bacterial membrane but with the purpose of injecting DNA. Phages take over the bacteria's genetic machinery in order to produce more phages, rather than bacteria. An interesting footnote in this passage is that phages are also the basis of genetic engineering: Geneticists insert certain genes into phages, prompting the bacteria to manufacture those genes.

Resistant Bacterium Modifies its Genome to Cause Infection

Scientists have identified a virulence region never seen before in the genome of *Enterococcus faecalis*—a leading cause of bacterial infection among hospital patients. This bacterium lives peacefully in the human gut, but it also thrives on wounds and burns. The researchers identified a group of genes that may contribute to the bacterium's transformation from a harmless gut homebody to a menacing invader.

A growing concern among doctors is the potential of *E. faecalis* to pass on antibiotic-resistant genes to other strains and pathogens. Hospital wards are by their nature a good place for the transfer of virulence factors among bacteria.

The discovery of the region, known as a pathogenicity island, could help in the fight against bacterial infections. The island may contain "useful markers for detecting unusually pathogenic strains," the researchers say. Armed with information about these harmful strains, hospitals could establish controls to prevent the spread of infection within their wards.

Michael S. Gilmore of the University of Oklahoma Health Sciences Centre and colleagues compared the genomes of several *E. faecalis* strains, including the strain responsible for an infectious outbreak twenty years ago. They also studied the first strain known to be resistant to the drug vancomycin—an antibiotic of last resort. All infectious strains contained a 150,000 base-pair region not found in the non-infectious strains.

The researchers also report that subtle variations within the structures of the pathogenicity island "enable strains with the element to modulate virulence." These variations, they note, occur at a high frequency.

"The danger of the organism is that it can be genetically flexible in persons with compromised immune systems," says Gary M. Dunny of the University of Minnesota in Minneapolis. The bacterium's virulence, he adds, probably involves other genetic elements outside the pathogenicity island.

The pathogenicity island contains genes coding for proteins thought or known to be important for virulence or survival of the bacterium. The region also codes for the surface protein Esp, which promotes bacterial attachment to surfaces and biofilm growth. Newly identified gene clusters in the island may be targets for new anti-bacterial therapies, the researchers report in *Nature*.

Novel Screen for Antibiotics uses Bacteria near Death

To meet the growing need for improved antibiotics, scientists a few years ago began to develop a drug-screening method based on arrays of genetically modified bacteria. That method, they now report, can be used to rapidly identify compounds that kill bacteria by inhibiting targets inside the bacterial cell.

In the method, researchers engineer bacteria to produce very low levels of an essential protein. Just barely alive, these bacteria are placed on an array and systematically exposed to a library of compounds. Compounds that kill more than one bacterium are strong drug candidates.

Because the array contains bacteria with related genetic flaws, the method reveals compounds that inhibit several targets. These will be more robust against anti-bacterial resistance than drugs that inhibit a single target protein.

"This method is a way to screen lots of compounds and look for very specific anti-microbial activity and lowered potential for acquired bacterial resistance," says David L. Pompliano, of GlaxoSmithKline in Collegeville, Pennsylvania, who led the study. The research, which included testing the strategy on *E. coli* bacteria, was done at DuPont Pharmaceuticals.

The most time-consuming part of the process is to work out the growth conditions for the bacterial strains. Some protein targets are expressed at levels far greater than what the bacterium needs to survive, and the goal is to get the expression down to the minimum level required for life.

"The beauty of this strategy is that you don't need to know the function of the target protein to do the study," says Pompliano. "So you can go to bacterial genomes and pick out a protein target found in many bacteria."

Antibiotics from a Microbe

The Genome of an Industrial Organism: Japanese researchers have reported the genome sequence of *Streptomyces avermitilis,* a free-living bacterium that grows in the soil. The bacterium is commercially important for human and veterinary medicine because it is one of the main producers of antibiotics. The sequencing revealed many genes involved in producing natural antibiotics that reside in clusters near both ends of the linear-shaped chromosome. These gene clusters interest researchers because understanding the antibiotic-producing mechanism could help generate novel drugs and slow down antibiotic resistance.

For bacterial standards, *S. avermitilis* has an unusually large genome of at least 8.7 million base pairs that consists of one linear chromosome. The team of researchers headed by Satoshi Omura, of the Kitasato Institute, in Tokyo, Japan, analysed 99 per cent of the genome using the whole-genome shotgun sequencing method. Altogether, they found at least 7,600 genes. Researchers have attributed the large genome size to extremely variable environmental factors the bacterium has to face as a free-living organism in the soil.

"The most interesting property of Streptomyces is its ability to produce secondary metabolites including antibiotics and bioactive compounds valued in human and veterinary medicine, agriculture, and unique biochemical tools," Omura—a specialist in the chemistry of natural products—and colleagues write in Proceedings of the National Academy of Sciences. Secondary metabolites are compounds with anti-bacterial, anti-fungal, and anti-viral properties and encompass a large variety of antibiotics. Researchers believe that these metabolites are produced by the bacterium at the end of its vegetative growth.

The scientists identified 25 kinds of gene clusters related to secondary metabolites. Even though these clusters are spread throughout the chromosome, the researchers were surprised to find about half of them located near both ends of the chromosome. "Some of the secondary metabolite clusters might have been horizontally transferred from donor microorganisms in the past," the researchers suggest.

Out of the 25 gene clusters, eight were found to produce compounds called avermectins. Avermectins are among the most important drugs for treating parasitic infections of livestock and humans, according to the researchers. The bacterial strain used in the study is *S. avermitilis* ATCC31267.

Variations in the penA gene of Neisseria meningitidis Increases Penicillin Resistance: In many countries, penicillin is no longer the invincible treatment for infections like meningitis and septicaemia. Now, European researchers have linked the declining efficacy of the antibiotic to mutations in the *penA* gene in *Neisseria meningitidis*, the bacterium behind these meningococcal infections.

N. meningitidis infects only humans, causing extremely

invasive infections that require urgent medical care. The treatment of choice has been penicillin. But reports of reduced susceptibility to penicillin are on the rise—and the cause has been linked to structural changes in the penicillin-binding protein (PBP2), which is encoded by the *penA* gene.

Aude Antignac and colleagues at Institut Pasteur in Paris, France, collaborated with researchers from Greece and the Czech Republic to study the relationship between polymorphisms in the bacterium's *penA* gene and the resistance to penicillin. Their results are published in The Journal of Antimicrobial Chemotherapy.

The team analysed strains of *N. meningitidis* from various genetic lineages that have been isolated in several countries over the last decade. They found strains with reduced susceptibility to penicillin (pen) that show a high degree of polymorphism, which frequently alters the amino acid sequence of PBP2.

But PBP2 was the only one of the bacterium's penicillin-binding proteins that contained variations. This finding suggests that pen strains are still evolving. If more of the bacterium's penicillin-binding proteins acquire mutations, full-blown resistance could come in the near future, much as it has with *Streptococcus pneumoniae* and other infectious bacteria.

Since reduced response to penicillin in *N. meningitidis* is directly linked to mutations of the *penA* gene, molecular methods of surveillance for these mutations may prove an invaluable tool for targeting penicillin-resistant strains. This will allow doctors to implement swift and effective treatment for dangerous meningococcal infections.

Marine Microbial Biodiversity

Microorganisms shape the biosphere of the earth by their spectacular geo-chemical activities. Microbes form an integral part of ecosystem and are represented by groups such as eubacteria, archaea, fungi, unicellular algae, and protozoa, etc., which are divided among three domains of life. The later three are eucaryotes-single cells or simple colonies of cells whose nuclei are wrapped in membranes. Among the remainder-called prokaryotes-genetic and physiological differences defined two further domains: the bacteria and the archaea. Microbes have proven to be a rich source of novel compounds with diverse distribution and biological activities (Davis 1999; Demain 1998 and 1999; Imada and Hotta 1992; Bernen *et al.* 1997). The oceans contributes more then 70 per cent of the earth surface of which about 60 per cent is covered by water more than 2000 mts deep. Paradoxically the oceans represent earth's least studied ecosystem especially for exploration of microbial diversity

The need to understand the role of the oceans on earth has never been more important than now because of the threats to our planet, particularly from climate change and ocean acidification. It was long thought that the ocean harbours far fewer species than the land. Of the 1.8 million species identified and named only around 2,75,000, or about 15 per cent, are of marine origin. Yet, as only a fraction of living species are known to science, these numbers fall far short of the world's real diversity. Marine microbiology encompasses the study of microbes that live not only in the superficial marine waters but also the abyssal depth from the coastal to the offshore regions, from the general oceans to the special niches like the blue waters of the tropical coral reefs ecosystems and to the black smokers of hot thermal vents. Though the oceans take up major portion of the living space on Earth, they have been poorly studied because their most parts are remote and extraordinarily hard to access. Only about 7 per cent of it has been sampled for its microbial diversity, microflora, microfaunal diversity, and also for other research and developmental activities. When it comes to mind-boggling numbers, marine microbiologist's can give anyone a run for their money as these microbial wealth is of industrial and academic interest. The oceans are brimming with more than 3×10^{28} bacteria-that's about hundred million times more cells than there are stars in the visible universe. But we have little idea what most of them are doing. The role that the ocean plays in structuring the earth's climate is largely driven by microorganisms (Copley 2002). Only a tiny fraction of ocean-dwelling microorganisms have ever been cultured in the laboratory. Hence, a better understanding of the microbes may be crucial to revealing how our planet works.

Today we are a witness to a revolution in microbial diversity assessment. Just as the first microscopes unveiled an unseen microbial world, the use of molecular techniques to enumerate different kinds and numbers of single-cell organisms has forever changed perceptions of the natural world. Microbial diversity has, now, been found to be at least 100-1000 times greater than the previous estimates based upon cultivation-dependent surveys (Pace 1997). As reviewed by Norse (1995), understanding marine biodiversity is very important for a number of reasons. One could argue that marine biodiversity has innate importance, as life has value on its own. If one considers how long life on earth has been in existence (some 3 billion years), it is apparent that the earth has been working (through evolution) on the present-day biodiversity for a long time. It is possible that we have little right deciding the fate of biodiversity in the first place. All ethics aside, however, the importance of marine biodiversity falls into following few categories: Adaptation, Food, Raw materials, Climate and Knowledge.

In marine environments, bacteria and fungi play a major role in recycling nutrients by converting organic detritus to a more usable form. Oceans because of their vastness and great depth are treated to be as the best dumping site of garbage's, which may be organic, inorganic or even radioactive in nature. It is a matter of great concern how these wastes are being recycled in the marine ecosystem. In this connection, the reports of Scholes *et al.* (1964) suggested the role of microorganism in recycling various organic and inorganic matters being one abiogentic factors such as, oxidation, light and salt played a very minor role. It was also suggested that there are at least two ways in which microorganisms are concerned in biodeterioration: (i) They may attack the material chemically or (ii) They may only foul the surfaces, where bacteria initially concentrated in the production of the primary film paving the way for later colonisation by other organisms. These chemicals which was synthesised by the microorganisms for biodeterioration of various metals, non-metals and other organic or inorganic matters are considered to be various types of enzyme, which played a major role in maintaining the bio-geochemical cycle in the sea. Communities of Bacteria, Archaea and Protists in marine environments are the primary catalysts of energy transformation and are responsible for more than 98 per cent of the carbon and nitrogen cycling (Atlas and Bartha 1993). An estimated 3.6×10^{30} microbial cells with cellular carbon of $\sim 3 \times 10^{17}$ grams may account for more than 90 per cent of the total oceanic biomass and 98 per cent of primary production (Atlas and Barthal 1993; Whitman, *et al.* 1998). Stable isotope studies reveal that for more than 3 billion years, these microscopic factories—initially anaerobic and later aerobic developed an enormous metabolic repertoire to cope with Earth's dynamic environment and mediated biogeochemical processes that shaped the planetary habitability (Des Marais and Walter 1999).

The oceans world wide are teeming with microscopic and macroscopic life forms. Due to higher level of nutrients, the microbial population is denser in the surface water along coasts, in shallow seas and along the equator than elsewhere. But the deep oceans are filled with life of great diversity, most of which depends on the rain of carbon from above. Rich chemosynthetic microbial communities thrive at deep-sea hydrothermal vents (Jannasch and Mottl 1985). Abundant Archaea populate oceanic mid-waters (Karner *et al.* 2001). Very large populations of phytoplankton including diatoms, dinoflagellates, picoflagellates and cyanobacteria are the primary catalysts in carbon fixation (Garcia-Pichel 2000) and cycling of nitrogen (Capone *et al.* 1997) and form the base of the traditional marine food web. The number of bacteriophage and viruses may be 100-fold higher. With such enormous population, the accumulation of mutations should lead to very high levels of genetic diversity and phenotypic variation. Yet, traditional microbiological methods have described only 30,000 protists (Corliss 1984; Lee *et al.* 1985; D.J. Patterson 1999) and fewer than 5000 kinds of prokaryotes (Guerrero 2001).

Based upon the traditional methods, the number of recognised microbial OTUs (Operational Taxonomic Units) is almost trivial when compared to estimates of 10^6 to 10^8 species for marine fauna. In contrast to animals and plants, the morphology of microorganisms is generally too simple to serve as a basis for a sound classification and to allow for reliable identification. Therefore, traditional microbiological methods have become outdated. They involve a long strenuous task of isolating microorganisms in pure cultures (or defined co-cultures) followed by identifying them on the basis of their phenotypic characters such as morphology, staining properties, metabolic capabilities, multiple physiological and biochemical traits. Nowadays, culture-independent descriptions of microbes are much preferred.

They rely upon comparisons of homologous genes (phylotypes) and a much richer microbial diversity in the marine environments have been revealed, while traditional microbiology had failed to culture more than 99.9 per cent of these newly discovered "phylotypes". Therefore the detailed analysis of microbial diversity within an environment can be divided into two broad categories: culture dependent studies and culture independent studies (Juck *et al.*, 2000). In cultured dependent methods, bacteria are isolated from environmental samples with culture medium. But the biggest drawback in exploring bacterial biodiversity is the issue of viable but non-cultivated organisms. Diversities in bacterial communities are normally determined by phenotypic characterisation of isolated strains. But the problem is that phenotypic methods can be used only on bacteria which can be isolated and cultured.

Though many advances have been made in microbiological culture techniques, it is still not possible to grow a majority of bacterial species using the standard laboratory culture techniques. Conventional characterisation of microbiology strains therefore has been subjected to debate as it is dependent on the ability of the strains to grow under specific environmental conditions (Bakonyi *et al.*, 2003). These types of classic microbiological methods are indirect and produce artificial changes in the microbial community structure. So, most bacteria will be excluded when phenotypic diversity is estimated. The isolated bacteria may account for only a minor proportion of the total bacterial diversity in the environment, while our knowledge about the dominant part is very scant. This has led to the current uncertainty regarding the true extent of bacterial life and leads to think of some modern techniques to study the wholesome microbial diversity.

Modern Techniques

Amazing advances in microbiology over the past 50 years force us to think in terms of ever shifting boundaries between the known and the unknown about single-cell organisms. In the late 60s, microbiologists had lost hope of constructing a robust natural system for microbial taxa. New molecular techniques developed during the 1970s opened pathways for establishing microbial phylogenetic relationships that were still unknown. The hallmark achievement in this field changed the entire scenario when Carl R. Woese and co-workers began to investigate the phylogeny of prokaryotes by cataloguing 16S rRNA. Using this 16S rRNA approach, they clearly demonstrated that life can be divided into three major lines of descent, now named the domains Archaea, Bacteria and Eucarya (Woese *et al.*, 1990). Any approach to identify specific microbial populations without cultivation directly in their natural environments became revolutionary since it completely changed the character of microbiology and closed the methodological gap existing in comparison with botany and zoology. As per Karl (2002) by most accounts, we can culture fewer than 10 per cent (by number) of the microbial inhabitants of the sea. So existing ideas of marine ecology must be flexible and accommodating to change. Both the stakes and the level of excitement in microbiological oceanography are at an all-time high, and in the field is poised for even more significant discoveries in the near future. Modern technologies like epifluorescence microscopy, electron microscopy, analysis of fatty acid patterens, respiratory isoprenoid quinines *(Busseetal,* 1996) and molecular techniques such as automated fluorescence cell sorting, combination of microautoradiography and fluorescence *in situ* hybridisation (FISH), flow cytometry and bioinformatics tools such as genomics and proteomics, etc. have demonstrated the great abundance and diversity of microbial life forms in the oceans.

DNA sequencing of environmental genomes (metagenomics) provide evidence of hitherto unrecognised physiological categories among the planktonic microbes. Sequence comparisons of PCR products (PCR amplicons) that target phylogenetically conserved regions of RNA coding genes, demonstrated that microbial diversity ranges from 10^5 to greater than 10^7 kinds of organisms (Pace 1997). An advanced method, terminal restriction fragment length polymorphism (T-RFLP) analysis, measures the size of terminal restriction fragments from a PCR- amplified marker.

It combines at least three technologies, including comparative genomics/RFLP,PCR & electrophoresis. Another technique, denaturing gradient gel electrophoresis (DGGE) and its cousin TGGE (Thermal GGE) is a method by which fragments of DNA of the same length but different sequence can be resolved electrophoretically (Muyzer 1999). However despite the impact of new information provided by the molecular biology toolbox, traditional techniques must not be abandoned since it is within this context that our understanding of marine microbial ecology has developed.

The hallmark of microbial diversity is biochemical innovation that single-gene studies cannot fully describe. Within the next few years, molecular biology will allow us to incorporate a definition for functional capacity or inducible phenotype in descriptions of microbial diversity (Zehr and Ward 2002; Steward *et al.* 2004). Microbiologists have now been able to identify the occurrence of a particular functional or structural gene and exploit it as a marker of diversity within an isolate or for members of a naturally occurring microbial population. In a similar manner, post genomic technology can measure gene expression patterns as a means to differentiate between "kinds" of microorganisms.

As a direct consequence of increased activity in marine metagenomics, the combination of high-throughput DNA sequencing, expression profiling and proteomics can describe new traits, novel functions and unusual enzymes in microbial populations. In some cases, entirely new phyla with novel functions are being discovered (Beja *et al.* 2002). These help in understanding the evolution of life and lead to sounder descriptions of new communities and species. Sequencing data will also be wedded to newly emerging molecular assays that incorporate automated sampling technologies and which would lead to finer temporal and spatial resolution of molecular diversity. If advances in genome technology and bioinformatics continue on the current trajectory, sequence scans of entire genomes or communities of genomes (Venter *et al.* 2004) coupled with high-throughput gene expression or proteomic profiles may become the standard for defining diversity and monitoring distribution patterns for microbial species.

However, no single tool allows definitive assessment of the bacterial diversity. Therefore, the use of polyphasic approach involving a combination of molecular biology techniques, microbiological methods and geochemical techniques or microsensors (Ramsing *et al.*, 1996; Teske *et al.*, 1996) is necessary to obtain a better understanding of the interaction between the microorganisms and their natural environment.

Research interest in microbial biodiversity over the past

25 years has increased markedly as microbiologists have become interested in its significance in various ecological processes and biotechnological (industrial, pharmaceutical and agricultural) applications. Microbes provide a variety of products (goods) and functions (services) which are essential to mankind's well being, including their role in almost all biogeochemical processes that sustain the biosphere such as in biogeochemical cycling of carbon, nitrogen, sulphur and phosphate Indeed, key aspects of the carbon and nitrogen cycles are carried out solely by bacteria, *i.e.* anaerobic fermentation of carbohydrates and fixation of atmospheric nitrogen), production of food and natural substances, assimilation of wastes and the regulation of world's climate. As such, it is important to determine various fact in the ecosystem, viz.

i. The types of bacteria present in a particular ecosystem,
ii. The role they play in the functioning of that system and
iii. The effect of natural calamities like Tsunami on them,
iv. Whether there are any endangered species present which needs conservation and
v. The effects that anthropogenic forces (particularly pollution) exert on them.

High levels of nutrients due to coastal pollution causes increased productivity which in turn can cause oxygen depletion in the water. Indeed, it has already been shown that sewage sludge dumping off the coast of New Jersey has impacted the benthic microbial ecology of the site (Hill *et al.* 1993).

Habitat distribution of the huge marine microorganisms is categorised into seven basic kinds; such as: 1. Neustonic (air sea interface), 2. Planktonic (bacterioplankton-0.2 to 2 μm or picoplankton), 3. Nekton (large living animals), 4. Seston (animal produced organic debris), 5. Epibiotic, endobiotic (symbiotic), 6. Pelagic (water borne habitats) and 7. Benthos (sea-sediment interface). To develop a description of microbial biodiversity in the oceans, the Census of Marine Life (CoML) had proposed an International Census of Marine Microbes (ICoMM) in 2004. It had embraced a worldwide strategy to explore the diversity and distribution patterns of all kinds of single-cell organisms in marine environments. Because we know so little about the limits of microbial diversity or whether biogeographical distribution patterns exist for microorganisms, major advances are expected to occur by 2010 albeit complete descriptions may require further decades of research.

In the rapidly emerging field of marine biotechnology, much of its success relies on investigation, characterisation and maintenance of marine biodiversity. Similarly, it will only be possible to accurately determine the structure and function of marine ecosystems if the types of microorganisms colonising a particular habitat and their metabolic processes are determined. For this purpose, bio prospecting (where a relatively small quantity of a large number of organisms are taken for general primary screening purposes) and bio-harvesting (where large quantities of a targeted organism are collected) are now popularly used.

Commercial Application

The marine environment is proving to be a valuable source of novel bioactive compounds with anti-bacterial, anti-viral, anti-cancer and anti-fouling properties (Colwell and Hill 1992). Research in varied aspects of marine biotechnology has thrown up a number of technologies. Nowadays, marine microbes are of varied commercial applications and used in the production of biopolymers, food and food industries, the production of enzymes, pharmaceutical industries, as biological response modifiers and environmental biotechnology (Vandamme 1989; Watanabe *et al.* 1996).

Biopolymers and surfactants have tremendous application value in food and petroleum industries. The microbial colonisation of the surface appears to depend on the ability of the microbe to adhere to the surfaces. A marine isolate of *Adnetobacter* was found to produce a surfactant "EMULSAN" which was later commercialised (Shete *et al.* 2004). The product was successfully marketed as a chelating agent for oil tankers and oil storage tanks. This compound has also the ability to complex with uranium, which may find application in recovery of uranium from waste material. These polymers are also useful in enhanced oil recovery (EOR). Nowadays, biopolymers like the polyester poly-â-hydroxybutyrate (PHB) are easily extracted from some extreme halophilic bacteria (Vandamme *et al.* 2003). PHB possesses the important properties of thermoplasticity, biodegradability and biocompatibility which have attracted considerable commercial attraction.

Marine microbes like *Spirulina* sp. and yeasts are now successfully used as sources of single cell protein (SCP). Single cell oils (SCO), the edible oils extracted from microorganisms, are best produced by various species of marine yeasts and fungi. Several key algae and bacterial sp. are also able to produce high levels of nutritionally important long-chain (LC) polyunsaturated fatty acids (PUFA). Oleaginous bacteria associated with marine sponges, *e.g. Bacillus subtilis* in *Aurora globostellata* and *Pseudomonas* sp. in *Heteronema erecta* have been found to produce total lipids 16.9 and 31.7 per cent, respectively, of their dry biomass (Pattnaik and Sree, 2005). Single cell oils are derived lipids mostly of triglycerol types which accumulate in the biological membranes as well as some macro-storage molecules in the cytoplasm of certain micro organisms.

Marine microorganisms also provide the basis for development of biosensors, bio-indicators, and diagnostic devices for medicine, aquaculture and environmental monitoring. Biomonitor that holds great promise is the gene probe, which can be used to identify organisms that pose health hazards or may be useful in research. Specific gene probes can be employed, for example, to detect human pathogens in seafood (Ramaiha *et al.* 1992).

There is an increased interest to switch to bio-pigments in

food, beverage, cosmetics, pharmaceuticals and fine chemical sectors than their synthetic chemical counter parts. Microbes and algae of varied groups are used for bio-pigment production. Among the carotenoids and xanthophylls, beta-carotenes and lycopene, leutin, zeaxanthin, canthoxanthin, rhodoxanthin and asthaxanthin attract full attention of industrial microbiologists (Goodwin 1980; Vandamme 1989-1992; Nelis *et al.* 1991). The industrial fermentation of zeaxanthin from a marine *Flavobacterium* sp and canthaxanthin from the marine bacterium *Rhodococcus maris* are reported by Nelis *et al.* (1991).

Marine microbes also play a key role in removing alkyl suphides, nitrate, nitrite in the wood-pulping industry (Visscher *et al.* 1993), degradation of soil fumigants, such as methyl bromide and 1,2 dichloropropane (Arp *et al.* 1993). Several strains of purple-non-sulphur marine bacteria are responsible for mineralisation of chlorinated aromatic compound, e.g. 2-chlorobenzoic acid, 2-chlorophenol or phenol (Kohring *et al.* 1993).Enzymes produced by marine bacteria are important in biotechnology due to their range of unusual properties. Some are salt resistant, a characteristic that is often advantageous in industrial processes (detergents and industrial cleaning application). Another research has demonstrated the presence in algae of unique haloperoxidases (enzyme catalysing the incorporation of halogen into the metabolites). These enzymes could become valuable products, as halogination is an important process in chemical industries (Taylor *et al.* 1993). Several fungi isolated from the coastal area of India have been reported to produce lactase and other lignocellulolytic modifying enzymes (Honda *et al.* 1985).

Marine enzymes are used to study the ion channels (like saxitoxin and tetrodotoxin as Na^+ channel blockers, brevitoxins Na^+ channel activators) (Bongiorni *et al.* 1996), biosensors for detecting environmental pollution (Condee *et al.* 1991), pathogen detection and disease diagnosic in aquaculture. Marine enzymes have also been applied in preparation of organic fertilizer, pollution control like waste water treatment

(Kawazoe 1993), degradation of plastic (Bongiorni *et al.* 1996), degradation of toxic pollutants, e.g. phenol and its methylated derivatives (Nordlund *et al.1993)* aromatic hydrocarbon and alkanes (Sotsky *et al.1993)* and bioremediation of fossil fuels (Betts 1993).

Many bioactive substances extracted from the marine microorganisms have also been found their application in pharmaceutical industries. They include anti-microbial compounds, anti- inflammatory agents, hormone-like materials, biological response modifiers, biosurfactants,, etc. A number of bacterial species that produce various anti-microbial compounds have been found in intimate association with sponges from the Caribbean coral reefs (Colwell 1997). The marine bacterium *Candidatus* sp. has recently had been found out to be a source of anti-cancer drug Bryostatin-1 (Haygood *et al.* 2002). Recently, *Vibrio vulnificus* has been found out to secrete a broad-specificity metalloprotease, which is capable of interfering with blood homeostasis through prothrombin activation and fibrinolysis (Chang *et al.* 2005).

Though limited information exists on threats faced by bacteria in the marine environment, many marine bacteria such as vibrios, members of enterobacteriaceae have been proved to produce toxins that are potentially dangerous to animal and human health. Certain cyanobacteria species and filamentous bacteria that form plankton blooms in estuarine, lagoonal and coastal waters are of importance in causing nuisance, as are many pathogenic bacteria. Halophilic marine vibrios such as *Vibrio vulnificus* have been reported to cause gastroenteritis, sepsis, cellulitis leading to necrotising soft tissue infection after exposure to seawater or consumption of raw seafood (Lowrie *et al.* 2003)). Therefore, the identification, pathogenesis and treatment of these potential marine pathogens becomes a major need.

Bibliography

Alfoldi, L.: *Hydrogeologic Considerations for in situ Bioremediation*, Butterworth-Heinemann Publishers, Stoneham, 1991.

Anderson, T. A. and Coats, J. R.: *Bioremediation Through Rhizosphere Technology*, American Chemical Society, Washington, 1994.

Antebi, Elizabeth, and David Fishlock: *Biotechnology, Strategies for Life*, Massachusetts, Cambridge, 1986.

Bailey, Ronald: *Ecoscam, the False Prophets of Ecological Apocalypse*, St. Martin's Press, New York, 1993.

Barley, S., J. Freeman and R. Hybels: *Strategic Alliances in Commercial Biotechnology*, Harvard Business School Press, Boston, 1992.

Batra, R. and W. Klassen: *Public Perception of Biotechnology*, Agricultural Research Institute, Maryland, 1987.

Biotechnology Industry Organization (BIO): *Agriculture and the New Biology*, The Association, Washington, 1987.

Biotechnology Research Subcommittee: *Committee in Fundamental Science and Research*, National Science and Technology Council, Washington, 1995.

Blakely, E.: *The Economic Development Potentials of California's Biotech Industry, Biotech Industry Research Group, Institute of Urban and Regional Development*, University of California, Berkeley, 1989.

Blakely, Edward J. and Kelvin W. Willoughby: *Institute of Urban and Regional Development, Biotechnology Industry Research Group,* University of California, Berkeley, 1990.

Blum, Deborah: *The Monkey Wars,* Oxford University Press, London, 1994.

Booz, Allen & Hamilton: *Foreign Investment in Bay Area Bioscience,* Bay Area Bioscience Center, Oakland, 1991.

Center for Agricultural Molecular Biology: *Cook College, New Jersey Agricultural Experiment Station, Rutgers,* The State University, New Brunswick, 1991.

Chong, Ken P. and Smith, John W.: *Mechanics of Oil Shale, London, England and New York,* Elsevier Applied Science Publishers, New York, 1984.

Cirpka, O. and Bisch, G.: *Technical Design of the Aquifer Model and Influence of Spatial Variability,* Presented at the Third International Symposium on In-Situ and On-Site Bioreclamation, California, 1995.

Committee on In Situ Bioremediation Water Science and Technology Board Commission on Engineering and Technical Systems National Research Council: *In Situ Bioremediation, When does it Work,* National Academy Press, Washington, 1993.

Committee on Scientific Evaluation of the Introduction of Genetically Modified Microorganisms and Plants into the Environment: *Field Testing Genetically Modified Organisms, Framework for Decisions,* National Academy Press, Washington, 1989.

Comstock, G.: *What Obligations have Scientists to Transgenic Animals, College Station, Center for Biotechnology Policy and Ethics,* Texas A&M University, *New Jersey,* 1992.

Cooney, J. J. and Gadd, G. M.: *Metals and Microorganisms, Journal of Industrial Microbiology,* Washington, 1995.

Crawford, Ronald L. and Don L.: *Crawford, Bioremediation*

Principles and Applications, Cambridge University Press, New York, 1996.

Dale, Jeremy W.: *Molecular Genetics of Bacteria*, John Wiley and Sons, New York, 1989.

De Lorenzo, V.: *Genetic Engineering Strategies for Environmental Applications*, Current Opinion in Biotechnology, California, 1992.

Drexler, K. Nanosystems: *Molecular Machinery, Manufacturing, and Computation*, John Wiley and Sons, New York, 1992.

Economidis, I. and Palmisano A.: *Environmental Biotechnology, EC-US Task Force on Biotechnology Research*, Brussels, Belgium, 1995.

Ehrlich, H. L.: *Geomicrobiology*, Marcel Dekker, New York, 1990.

Esser, K.: *Plasmids of Eukaryotes, Fundamentals and Applications*, Springer-Verlag Berlin Heidelberg, Germany, 1986.

Fowle, John R.: *Application of Biotechnology, Environmental and Policy Issues*, Westview Press, Colorado, 1987.

Francis, A. J.: *Microbial Dissolution and Stabilization of Toxic Metals and Radionuclides in Mixed Wastes*, Oxford, New York, 1990.

Frankenberger, W. T. and Benson S. M.: *Selenium in the Environment*, Marcel Dekker, New York, 1994.

Friedman, D.: *The New Economy Project Final Report*, The New Economy Team, UCLA, 1994.

Gendel, Steven: *Agricultural Bioethics, Implications of Agricultural Biotechnology*, Iowa State University Press, Ames, 1990.

Gibson, D. T. and Sayler, G. S.: *Scientific Foundations of Bioremediation, Current Status and Future Needs*, American Academy of Microbiology, Washington, 1992.

Ginzburg, Lev. R.: *Assessing Ecological Risks of Biotechnology*, Butterworth-Heinemann, Boston, 1991.

Goodsell, David S.:*The Machinery of Life*, Springer-Verlag, New York, 1993.

Grimes, D. J.: *Validation of Reliability and Predictability of Environmental Biotechnologies*, Workshop on Bioremediation, Japan, 1995.

Grocery Manufacturers of America: *Safe Foods Developed Through Biotechnology, Nutritious, Healthful*, Abundant and Good Tasting, 1993.

Hacking, Andrew J.: *Economic Aspects of Biotechnology*, Cambridge University Press, Cambridge, 1986.

Harkness, M. R., McDermott J. B. and Abramowicz D. A.: *In Situ Stimulation of Aerobic Degradation of PCB Biodegradation in Hudson River Sediments*, MacMillan, London, 1993.

Hettinger, N.: *Owning Varieties of Life, Biotechnology, Intellectual Property, and Environmental Ethics, Center for Biotechnology Policy and Ethics*, Texas A&M University, Texas, 1994.

Hettinger, N.: *Biotechnology Policy and Ethics*, A&M University, Texas, 1992.

Hinchee, R. E., Anderson D. B., Metting, F. B. and Sayles G. D.: *Applied Biotechnology for Site Remediation*, Lewis Publishers, Florida, 1994.

Hinchee, Robert E.: *Applied Biotechnology for Site Remediation*, CRC Press, Ann Arbor, 1994.

Hollenberg, C. P. and H. Sahm: *Microbial Genetic Engineering and Enzyme Technology*, Gustav Fischer, New York, 1987.

Kelly, Kevin: *Out of Control*, Addison-Wesley, Menlo Park, 1994.

Kenney, Martin *Biotechnology: The University-Industrial Complex*, Yale University Press, Connecticut, 1986.

Kimbrell, Andrew: *The Human Body Shop*, Harper/Collins, San Francisco, 1993.

King, R. B., Mihony, G. and Sheldon J. K.: *Practical Environmental Bioremediation*, Lewis Publishers, Florida, 1992.

Kloppenburg, Jack Ralph: *The Political Economy of Plant Biotechnology*, Cambridge University Press, Cambridge, 1988.

Koehler, Gus.: *Tools to Promote Economic Success*, California Research Bureau, Sacramento, 1995.

Kotkin, Joel and Levy, Stephen: *A Twenty-First Century Prospectus, Center for the New West, and Center for the Continuing Study of the California Economy*, California, 1996.

Krimsky, Sheldon: *Biotechnics and Society,The Rise of Industrial Genetics*, Praeger, New York, 1991.

Lacy, W., L. Busch, and L. Lacy: *Agricultural Biotechnology, Issues and Choices*, Purdue University Press, Indiana, 1991.

Lacy, W., L. Lacy and L. Busch: *Emerging Trends, Consequences, and Policy Issues in Agricultural Biotechnology*, Westview Press, Colorado, 1991.

Lee, V. and Yen T.F.: *Biopolymer Shields, in Applied Biotechnology for Site Remediation*, Lewis Publishers, Florida, 1994.

Levorsen, A. I.: *Geology of Petroleum, San Francisco*, W. H. Freeman and Company, California, 1967.

Lovley, D. R.: *Dissimilatory Metal Reduction*, Annual Review of Microbiology, 1993.

Lynch, J. M. and Hobbie J. E.: *Concepts and Applications in Microbial Ecology*, Blackwell Scientific Publications, England, 1988.

McKinsey Global Institute: *Manufacturing Productivity*, McKinsey and Company, Washington, 1993.

Metting, Jr., F. B.: *Soil Microbial Ecology, Applications in Agricultural and Environmental Management*, Marcel Dekker, New York, 1992.

Miller, Henry I., M.D : *Remarks for the California Legislature*

Assembly Committee on Agriculture, MacMillan, London, 1994.

Miller, R. V. and Poindexter J. S.: *Strategies and Mechanisms for Field Research in Environmental Bioremediation,* American Academy of Microbiology, Washington, 1994.

Mueller, J. G.: *Model Studies of Bioremediation Strategies and Monitoring Technologies in Proceedings, Environmental Biotechnology,* EC-US Task Force on Biotechnology Research, Belgium, 1995.

National Research Council (NRC): *Alternatives for Ground Water Cleanup,* National Academy Press, Washington, 1994.

Old, R. W. and S. B. Primrose: *Principles of Gene Manipulation, An Introduction to Genetic Engineering,* University of California Press, California, 1980.

Olson, Steve: *An Industry Comes of Age,* National Academy Press, Washington, 1986.

Organisation for Economic Cooperation and Development (OECD): *Biotechnology for a Clean Environment, Prevention, Detection, Remediation,* OECD, Paris, 1994.

Ouellette, Robert P. and Paul N.: *Cheremisinoff, Essentials of Biotechnology, Lancaster,* Technomic Publishing Company, Pennsylvania, 1985.

Panem, Sandra: *Biotechnology, Implications for Public Policy,* The Brookings Institution, Washington, 1985.

Powell, W. and P. Brantley: *Competitive Cooperation in Biotechnology, Learning Through Networks,* Harvard Business School Press, Boston, 1992.

Prentis, Steve: *Biotechnology, A New Industrial Revolution,* George Braziller, New York, 1984.

President's Council on Competitiveness: *Report on National Biotechnology Policy,* The Council, Washington, 1991.

Rissler, Jane, and Margaret Mellon: *Perils Amidst the Promise,*

Ecological Risks of Transgenic Crops in a Global Market, Union of Concerned Scientists, Cambridge, 1993.

Saaty, R. P. and Booth S. R.: *Cost Effectiveness of a Remediation Technology Field Tested at the Savannah River Integrated Demonstration Site,* Los Alamos National Laboratory, New Mexico, 1994.

Saunders, Venetia A. and Jon R. Saunders: *Microbial Genetics Applied to Biotechnology, Principles & Techniques of Gene Transfer & Manipulation,* Crook Helm Ltd., London, 1987.

Saxenian, A.: *Regional Advantage,* Harvard University Press, Boston, 1994.

Sayler, G. S., Fox, R. and Blackburn, J. W.: *Environmental Biotechnology for Waste Treatment,* Plenum Press, New York, 1991.

Sayler, G. S.: *Biodegradation Process Analysis, Molecular Application in Simulations and Environmental Verification in Proceedings, Environmental Biotechnology,* EC-US Task Force on Biotechnology Research, Belgium, 1995.

Sayler, Gary S.: *Environmental Biotechnology for Waste Treatment,* Plenum Press, New York, 1991.

Skladany, G. J. and Metting, F. B.: *Bioremediation of Contaminated Soil in Soil Microbial Ecology,* Marcel Dekker, New York, 1992.

Smith, George P.: *The New Biology, Law, Ethics and Biotechnology,* Plenum Press, New York, 1989.

Sussman, M.: *The Release of Genetically-engineered Micro-organisms,* Harcourt Brace Jovanovich, New York, 1988.

Thomas, J. M. and Ward, C. H.: *In Situ Biorestoration of Organic Contaminants in the Subsurface,* Environmental Science and Technology, 1989.

Thompson, P: *Animal Welfare and Animal Rights, Center for Biotechnology Polilcy and Ethics,* Texas A&M University, Texas, 1992.

Thompson, Paul B., Robert J. Matthews: *Ethics, Public Policy and Agriculture,* MacMillan, London, 1994.

Tunlid, A. and White, D. C.: *Biochemical Analysis of Biomass, Community Structure, Nutritional Status, and Metabolic Activity of Microbial Communities in Soil in Soil Biochemistry,* Marcel Dekker, New York, 1992.

Wijesinghe, A. M., Knapp, R. B. and Carman, L. M.: *Preliminary Feasibility and Cost Analysis of the In-Situ Microbial Filter Concept,* Lawrence Livermore National Laboratory, Livermore, California, 1992.

Index

A

B

C

E

F

G

P

R

S

❑❑❑